MAPPING HUMAN SECURITY CHALLENGES IN KASHMIR VALLEY

MAPPING HUMAN
SECURITY CHALLENGES
IN KASHMIR VALLEY

Mohd Aarif Rather

St. James's Studies in World Affairs

ACADEMICA PRESS
LONDON-WASHINGTON

Library of Congress Cataloging-in-Publication Data

Names: Rather, Mohd Aarif, author.
Title: Mapping human security challenges in Kashmir Valley / Mohd Aarif Rather.
Description: London ; Washington : Academica Press, [2018] | Includes bibliographical references and index.
Identifiers: LCCN 2018039963 | ISBN 9781680534726
Subjects: LCSH: Human security--India--Kashmir, Vale of. |Militarization--India--Kashmir, Vale of. | Civil-military relations--India--Kashmir, Vale of. | Kashmir, Vale of (India)—Social conditions. | Kashmir, Vale of (India)--Politics and government. | Jammu and Kashmir (India)--History--Autonomy and independence movements. |India--Foreign relations--Pakistan. | Pakistan--Foreign relations--India.
Classification: LCC JC599.I42 K36474 2018 | DDC 954.05/3--dc23
LC record available at https://lccn.loc.gov/2018039963

I dedicate this book to my Beloved Parents,
For their endless support, love and
Encouragement.

Contents

FIGURES, MAPS AND TABLES

APPENDIX

LIST OF ABBREVIATIONS

AFSPA	Armed Forces Special Powers Act
APDP	Association of Parents of Disappeared Persons
BPL	Below Poverty Line
BSF	Border Security Force
CHS	Commission on Human Security
DPJ	Democratic Party of Japan
GER	Gross Enrolment Ratio
ICT	Information, Communication and Technology
IPTK	International People's Tribunal on Human Rights and Justice in Indian Administered Kashmir
J and K	Jammu and Kashmir
JKHRC	Jammu and Kashmir Human Rights Commission
LDP	Liberal Democratic Party
LoC	Line of Control
NHRC	National Human Rights Commission
NGOs	Non-Governmental Organisations
PBRC	People's Basic Rights (protection) Committee
PDP	People's Democratic Party
PSA	Public Safety Act
PUDR	People's Union for Democratic Rights
PUCL	People's Union for Civil Liberties
UN	United Nations
UNCIP	United Nations Commission on India and Pakistan

UNDP United Nations Development Programme

UNESCO United Nations Economic and Social Council

UNICEF United Nations Children's Fund

UNRWA United Nations Relief and Works Agency

UNTFHS United Nations Trust Fund for Human Security

US United States

USA United States of America

WHO World Health Organisation

FOREWORD

The book titled "Mapping Human Security Challenges in Kashmir Valley" is an in-depth study carried out by the author who rightly deserves appreciation. There are vast emerging contemporary challenges of Human Security day in and day out, and these challenges become much more violent and challenging in context of conflict zone areas like the Kashmir Valley. The societies being dynamic in terms of political, social, cultural, economic, technological advancements, have to keep pace with the changing scenario. These developments pose greater threat to human security in terms of life, freedom, safety and the dignity – the basic inherent universal rights of the human beings. The author has comparatively and critically analysed the various perceptions regarding human security which includes its origin, nature, threat and role of nation-states in the promotion and protection of human personality.

The bilateral dispute between India and Pakistan over the Kashmir Valley have been analysed by author in an unbiased manner. The author has beautifully tried to explore the human security concept, its challenges, state and non-state intervention in protecting or marring the human security. Human security is an international phenomenon concerning all the states globally. However, the extent, intensity and magnitude of Human Security threats may vary from state to state, but it concerns one and all. For instance, environmental degradation, drug and human trafficking, militancy knows no boundaries. The time is ripe that the whole world has to shoulder a collective responsibility in this regard in order to attain the maximum good for human being and for human security, so as to claim a rightful place in the civilised nations.

The author has conducted a deep and thorough study related to Kashmir conflict and traced the roots of the conflict. The study limelights the reasons and continued existence of the conflict and the role of central and the state government in power at various time intervals. The history and official data stand witness to the fact that basic tenants of human security have always been in want in the Valley of Kashmir. Besides, the loss to human life, mental depression, infrastructural damage multi-dimensionally, economic stagnation, educational backwardness has added to miseries. The draconian colonial laws like AFSPA, PSA still rule this region without even being revived. All this has been limelighted by the author concretely and lucidly. The author has categorically shown a ray of hope through the propose oriented dialogue, along with the active role of United Nations and the demilitarisation process for resolving the pending Kashmir issue.

This piece of work will be valuable for all the students of international relations, scholars and academicians who want to pursue their career in this field. The author has done a commendable job and worth appreciating. I wish him good luck.

Dr. Aneeda Jan
Assistant Professor (Law)
DDE, University of Kashmir

ACKNOWLEDGEMENT

First and foremost, I humbly acknowledge the grace of Almighty Allah bestowed upon me for the completion of this work. Without his blessings, this achievement would not have been possible. This journey has not travelled in a vacuum and have been supported and encouraged by numerous people including my family members, relatives, well-wishers and friends.

I would like to offer my deep gratitude and reverence to my esteemed teacher and supervisor Dr. Kishore Jose (Central University of Gujarat) for guiding me in every possible manner. His attentive supervision, logical guidance and deep concern provide me the way of my work. I often received illumination, encouragement and inspiration from him. I would also like to express my sincerest gratitude to teaching staff members of the Centre of Security Studies for assisting me in my work. I also express my profound gratitude to the non- teaching, technical and library staff members of the University for providing me necessary help.

I dedicate everything to my parents. Their support, care and affection are beyond the realm of words and expression. They always prayed for my success and provide me with unfailing support and continuous encouragement throughout my years of study and through the process of researching and writing this work. Special thanks to my Daddy Jan who has been an everlasting inspiration for me.

My acknowledgement would be incomplete without thanking my elder brother Adv. Mohammad Nazir Ahmad for helping me in difficult moments and for sincere support and unceasing encouragement. I am also grateful to my other family members particular my Di (Dr. Aneeda Jan) and younger brother (Prince Zahoor) for their unwavering and unselfish love and support given to me at all times. My heartful thanks are due to my Sisters- A.G, Baji, Guddi di, Meema di and to Abaji and Mummy. I express my love to my sweet niece "Meras" whose innocence refreshed me during my difficult times.

I extend my thanks to all my friends particularly Sushma Parmar for her unforgettable support to boost my confidence throughout my work and pleasant company which will be ever remembered. I thank all my other teachers, friends and relatives who have supported me along the way.

And last but not the least, special mention goes to my joyful and lovely niece and nephews Basit, Rani, Anjuman, Zahid, Ayesha and Ibraheem for their love and dua. In their company, I forget all worries.

Mohd Aarif Rather

CHAPTER – 1
INTRODUCTION

The nature of international politics has undergone profound changes since the end of cold war. The end of the bipolar world order has marked with the emergence of a new set of challenges. The globalisation process accelerated by Information, Communication and Technology (ICT) has drastically transformed every sphere of human life in terms of time and space that unleashes opportunities as well as challenges to nation states. In fact, many issues such as environmental degradation, terrorism, inequality in income distribution etc., have now confronted across globe in our age that was either hitherto missing or less visible in the security debate. Also, the nature of wars and conflicts have changed from inter-state to intra-state on ethnic, religious etc, lines. These changes have raised new questions in the field of security studies which were, till then heavily focusing on nation-state centric security challenges. Thus, the traditional security understanding seems to be incapable of addressing many of these issues and, hence, attempts were made to incorporate such concerns within the ambit of Security Studies. The Conceptualisation of Human security approach has to be understood in the backdrop of such sweeping changes.

It is within this context that the notion of Human Security came into picture and achieved a priority of top concern at global level. The concept of Human Security aims to protect an individual from threats not only in terms of external aggression but also to provide safety from threats and disruptions emerging in due course of their daily lives and to assure their survival and development. Thus, the main concern of Human Security approach is the overall human development, along with the fulfilment of basic necessities of human beings. The human security approach contends that the challenges and threats to security transcend national defence as well as law and order, to encompass all issues including political, social and economic that guarantee a life free from fear and risk. As such, the individuals attained prominence in security discourse (Hussein, Gnisci and Wanjiru 2004). But, this could not have profound implication for the countries of global south. Many Asian countries like India, China, North Korea, Iran and Pakistan still embarked on military and nuclear programmes (Bracken 1999).

The Human Security concerns are more apparent in the regions with violent conflicts (conflict zones) or in the state of disturbing or chaotic environments. In this perspective, the valley of Kashmir offers an ample ground to examine the challenges of Human Security. The valley of Kashmir witnessed the uncertainty in terms of social, political and economic factors since its merger with India in 1947 and most particularly from 1989 with the inception of militancy. Besides, the Kashmir valley remained as a contentious issue between two nuclear powers i.e., India and Pakistan and a reason of escalating tension between them. With the due course of time, it evolved as an international dispute. Although, both countries suffered heavy economic and human losses but the worst sufferers were the people of the valley. The violent environment prevailing in the valley since early 1990's added further inadequacies and nuances to the local people in the valley. In this context, the present study attempts to map the Human Security challenges in the valley of Kashmir.

Human Security: Background and Overview

The genesis of the concept of Human Security can be sketched if not traced back in the 1970s and 1980s. In 1970's, a series of volumes have been issued by the Club of Rome on "world problematic" focussing on the web of problems troubling men of all nations like poverty, insecurity of employment, environmental degradation, economic disruptions and alienation of youth (Bajpai 2000). These problems arise out of complex global system and influence the individual's life and thus need a proper attention to overcome these insecurity dilemmas. Also in 1980s, the north south report published by the independent commission on International Development chaired by Willy Brandt urged the need of primacy to the human being. The report further commended the necessity to overcome mass misery, world hunger and alarming disparities between the living condition of rich and poor (Independent Commission on International Development Issues 1980). Further in 1991, the Stockholm incentive on global security and governance ushered the need for common global responsibility to focus on the threats such as under development, environmental degradation, excessive population growth and movement, and the lack of progress towards democracy (Government of Sweden 1991). These commissions and reports may be viewed as the precursors to Human Security thinking.

However, the term 'Human Security' evolved in 1994 with the publication of United Nations Development Report (UNDR) and is broadly associated with the security of people - their physical safety, their economic and social well-being, respect for their dignity and worth as human beings, and the protection of their human rights and fundamental freedoms. Besides, the UNDR also pointed out few aspects of Human Security domains such as economic, political, environment, community, food and health security. The broadening of Human Security concept led to paradigm shift in thinking about security/threat in modern times. Human Security embraces not only the threat of life in the course of war and conflict but also the political liberty and the economic opportunities as well (UNDP Report 1994).

The notion of Human Security thus appeared as an emerging paradigm to address the current as well as projected global vulnerabilities and thereby challenging the conventional perception of national security. The proponents of the concept argue that human being is the referent object of security concerns implying that the individuals prevail over the nations or state's importance in security matters (Kis 2010). The objective of Human Security is to safeguard the vital core of all human lives from critical pervasive threats, in a way that is consistent with long term human fulfilment. Human Security takes its shape from the human being: the vital core that is to be protected (Alkire 2003). The idea of Human security was embraced by some of the middle powers like Japan, Canada and Norway as a guiding principle of their foreign policies. After the 9/11 attacks and the War on Terror, the conception of human security has been clearly acknowledged as an option in the rhetoric of the major powers (Gomez 2012).

The concept of Human Security is broadly based on two notions: freedom from fear and freedom from want. The former view propounded by Canada stresses on the safety of people from both violent and non-violent threats, freedom from pervasive threats to people's rights, there safety or even theirs rights. The later view articulated by Japan focuses on the security designed to protect people from threats to their livelihood and dignity such as poverty, diseases and environmental degradation. These two countries have proclaimed it the guiding principle of their foreign policy (Tadjbakhsh and Chennoy 2007).

Broadly there are three schools of thought regarding the meaning and implications of the concept of Human Security. The first approach to Human Security called "Rights-Based Approach" focuses on legal rights and rule of law. It viewed that the main threat to Human Security lies in the denial of fundamental human rights including the right of national self-determination and the absence of rule of law. The second approach called "Safety of People Approach" (sometimes described as freedom from fear) is centred on the protection of people from violent threats i.e. war and also in viewing economic despair, social injustice and political instability. The third approach termed as "Sustainable Human Development Approach" (also referred as freedom from want) is associated with the United Nation's Development Report of 1994. The approach views Human Security in terms of economic, food, health, environmental, personal, community and political security. Thus, looking at the above mentioned approaches, it becomes apparent that the focus of attention and concern of Human Security is the individual, rather that the state. The figure 1.1 on the next page shows approaches of human security.

Figure: 1.1. Approaches of Human Security

Source: Own Compilation

Human Security is mainly viewed to confront with threats at international level. Although some regional approaches to Human Security were launched by a series of nations like UNESCO studies for Arab states (2005), East Asia (2004) and Central Asia (2006). From the Indian perspective, two Human Security projects were development with the aim to provide specific tools that enable policy makers to address Human Security challenges, one was in 2007- 2008 in India's northeast region regarding conflict and development. The other was in 2008-09 in Orissa on the growing naxalite insurgency. Both studies have been undertaken in the context of development and conflict relation, given the fact both areas are conflicted ridden (Acharya, Singh, Rajaretnam 2011).

Locating the Kashmir Conflict

After the partition of British India in 1947, one of the major regions namely Kashmir valley has been conflict prone area within national boundary as well as at the international level. This conflict is marked as one of the most complex situation wherein on the one hand two nuclear powers i.e., India and Pakistan are involved, while, on the other hand, the people of the valley of Kashmir developed alienation against the Indian federal rule (Schofield 2000). However, conventionally, Kashmir has been largely viewed from bringing peace through military intervention. But over the last six decades of independence, India suggests that development or underdevelopment of Kashmir persist or even intensify.

The valley of Kashmir witnessed militarisation for external defence as well as the use of the armed forces for domestic repression that has transformed the valley into a source of deep insecurity for the people which had largely undermined the democratic ethos. However, the valley observed an extensive militarisation in the early 1990's due to the rise of militancy which mainly evolved by the political failure on the part of Indian government. As against this, the Indian state responded by categorising the valley as a 'disturbed region' and administered it through a legal provision of special laws like the Disturbed Areas Act, Public Safety Act (PSA) and the Armed Forces Special Powers Act (AFSPA). These special provisions authorise the Indian state to deploy a massive apparatus of over 700,000 military, paramilitary and police personnel. Presently, the militant phase of the struggle has passed away as there are only few hundred militants active in the valley as per the state officials but still the security forces (military) are stationed with the mission of maintaining law and order and thereby creating insecurities in everyday life in the valley. The presence of high ratio of troops to civilians makes the valley as one of the most militarized places in the world and producing an everyday experience of embattlement, of living life under siege and under occupation (Hoffmana and Duschinski 2013).

Throughout the Indian rule, the valley of Kashmir has witnessed militarization, economic deprivation, subjugation and indiscriminate violence including the denial of democratic processes, massive human rights violations, the manipulation of elections as well as jailing of political leaders (Behera 2000, Bose 2003). Also, the patterns of continuous exclusion and marginalization have entrenched the people of the valley a collective feeling of alienation from the Indian government. As a result, Kashmiris have been using the language of self- determination as a way of expressing their collective will in accordance to decide their own political future. However, it is not enough to locate the problems of valley in one dimensional way but demands multiple directions. For instance, like of economics opportunity, persistent poverty, unemployment, industrial backwardness and so on has some way or the other way kept the valley in trap of underdevelopment.

Human Security in Kashmir: Identifying the Parameters

On the path of secure human life, there are numerous challenges of which Human Security concept identified and attempt to give adequate diagnosis. The central concern of human life has been absence of violence that otherwise imposes severe restriction. In this way, the emergence of militancy in late 1980s and early 1990s, led to massive human rights violations in the valley of Kashmir. Due to this violent conflict, thousands of civilians have died in regular clashes between separatist militants and security forces. The Srinagar based international people's tribunal on human rights and justice in Indian administered Kashmir and the association of parents of disappeared persons prepared a 354 page report on human rights violations. The report said that an estimate of 8000 persons had disappeared in the past two decades, in addition to the death of more than 7000 people. It cited disclosers of more than 6000 unknown, unmarked and mass graves (Gayathri 2012).

Besides, the presence of high ratio of Indian troops in the Kashmir valley have resulted in massive human rights abuses in the form of disappearances, custodial killings, torture, rape and molestation of women. These violations committed by the Indian forces have been well documented by several human rights organisations such as Amnesty international, Asia watch, People's Union for Democratic Rights (PUDR), People's Union for Civil Liberties (PUCL) and so on. From 1989 onwards, the valley of Kashmir had witnessed over more than 70,000 deaths and 119,789 arrests of civilians. Out of these arrests made by armed forces, 6989 civilians have been killed in the custodians (Tabassum 2012). Although, there is no state of emergency in the valley but still there are around seven lack Indian forces stationed in the state mainly in the valley (Mohiuddin 1997).

The valley of Kashmir witnessed the presence of large number of armed groups and is claimed as the world's most heavily militarised places. These armed forces are operating under some draconian laws like Armed Forces Special Powers Act (AFSPA) and Armed Forces Special Powers Act and Public Safety Act (PSA). The AFSPA was promulgated in the valley of Kashmir in 1990 by which the armed forces possess the special powers. These special powers permit the armed forces to arrest and enter any property without warrant. Also, they are authorised to fire upon or otherwise use force and even causing to the death of any person, if the security person believes it necessary for the maintenance of public order (Singh 2011). In addition, the PSA which is only applicable to the state of Jammu and Kashmir by which a person can be detained for the period of two years without any trail vis-a-vis the detaining authority does not disclose the facts of detention. Under this act, estimates of the number of detainees in the last two decades could range from 8000- 20000 (Ibid).

Under AFSPA, the armed forces find a free hand in the name of law and order and often became the perpetuators of human rights violations. According to the report prepared by two human rights groups namely Indian Administered Kashmir (IPTK) and Srinagar-based Association of Parents of Disappeared Persons (APDP), almost 500 individual perpetrators including 235 army personal, 123 paramilitary personal, 11 J and K police personal and 31 government backed militants were involved in human rights violations over the last two decades (Gayathri 2012). Quoting Duschinshki (2009), under AFSPA, Kashmiris have become permanent prisoners of the state and through intensive militarization of Kashmiri society, India administered Kashmir valley through a form civil military authoritarianism.

Apart from militarisation and human rights violations, the Kashmir valley over the past 24 years of violent conflict has suffered from severe economic setbacks. Every sector of economy has been affected, be it the tourism, handicraft production, the horticulture, medium and small-scale industries etc. The production of handicrafts in Kashmir valley has deteriorated during 1989-90 as Rs 150 crores from Rs 154 crores in 1988-98 and during the peak seasons of turmoil in summer 2008, 2009 and 2012, the production got further deteriorated (Ahamed & Yasmeen 2012). Similarly, the conflict has hit hard the tourism sector leading to the heavy loss of tourism appetite. The tourism sector as being the main source of income in the valley suffered a virtual shutdown during the years of turmoil as the number of tourists visiting the valley per year had gone down from around 7,000,000 in the pre-militancy days to a few thousand in the following years (Directorate of Handicrafts J & K Government 2012-13).

The presence of violent environment in the valley of Kashmir has affected the development by discouraging private investment and thereby pushing the economy towards stagnation. It had also created obstacles in effective implementation of the developmental policies initiated by public authorities. Also, the state is unable to generate its own resource mechanisms leading to further underdevelopment. In absence of any developed industrial setup, agricultural sector as being the major source of livelihood for the majority of the people also remained poor and underdeveloped. The lack of progress towards the agricultural sector is mainly because of unorganised marketing facilities and absence of export markets, along with the inefficiency of state government to implement such policies that would have boost this sector.

The conflict in the valley had also increased the poverty leading towards malnutrition and hunger, thereby, creating food security issues. According to the official estimates, the total estimated Below Poverty Line (BPL) population ratio in the valley is nearly 13 lack people i.e, 21.37 Percent in 2012. (J and K Economic Survey 2012-13). In addition to the poverty, the valley is facing from acute unemployment problems as there are 3, 21,526 unemployed youth in the year 2012-13 (Kathju 2013). This unemployment further resulted in un-meeting of normal social life pushing towards further poverty. Also, the valley lags behind in terms of the literacy rate as it ranks to the 30th position at the national level (Raj 2013).

The Kashmir valley faced the Human Security challenges so far as the health security of the valley is concerned. Over more than two decades of the violent conflict, the valley witnessed thousands of deaths and injuries leading to thousands of people as limbs and orphans. The estimates showed that that the conflict has caused around 100,000 as orphans (Mcgivering 2000). Also, the health infrastructure has been heavily damaged which was already inadequate further added to the miseries of the valley. The conflict had also severely costed upon the health of women and children vis-à-vis on the mental health of the people. On the other hand, the valley lacks proper mental care facilities as there is single government psychiatric hospital with little capacity to provide proper treatment to psychiatric patients (The Hindu 2011).

In addition to the health problems, the conflict had further costed heavy damage to be infrastructure of the valley. The basic infrastructure of the valley i.e. roads, communication system, regular power supply and drinking water facilities remained poor in Kashmir. Further, it is also estimate that from 1989 to 2002, over 1151 government buildings, 643 educational buildings, 11 hospitals, 337 bridges, 10729 private houses and 1053 shops have been gutted in some 5268 attacks on infrastructure (Mahapatra & Shekhwat 2008). Also, as per the survey findings, the road length per 100 km in Kashmiris is 35.71 km as against 104.64 km in the country. Comparatively, the communication facility is also inadequate as is evident that there is one post office for a 60 sq. area to the national average of 20 sq. km. The valley of Kashmir is also suffering from acute safe drinking water facilities as it is not available to 45% people in rural areas and 4% in urban area areas (Ahamed & Yasmeen 2012).

The entrenchment of conflict and violence in the valley of Kashmir has deeply affected the human resource. The human resource suffered enormously in J and K due to large displacement of Kashmiri Pandits, Sikhs from the valley as around 55000 Kashmiri Pandits families flee their homes and took shelter in Jammu and elsewhere in the country (Singh 2011). However, the migration of this community which changed their sociological and ecological setting had multiplicity of effects on their living conditions and survival (Arora and Dhingra 2005).

Review of Literature

The review of literature of the present work adopts a structured and thematic approach. It tries to analyse and highlight the existing literary works and to find the various gaps therein. Although, there are several studies on the conception and applications of Human Security by various scholars, academicians etc. However, there exists a limited study on locating the issues of the Kashmir valley through the prism of human security approach. Some of such works are reviewed in the following manner.

To understand the concept of Human Security the present work starts with the book "*Human Security: Concepts and Implications*" by Tadjbakhsh and Chennoy (2007) which mentions the pioneered work of Mahbubul Haq and Amratya Sen during 1990 when United Nations first came up with Human Development report. Historically, development has been seen in different way and different era, for instance, before the Second World War and more precisely First World War, development has been seen as a trickle down approach, still some believe that development can be achieved by trickle down approach. Thus, it means that economic wealth of nations can be achieved through rigorous industrialisation and entrepreneurship without bothering about distributional aspect of income. But the Second World War and great depression during 1929 in western countries mid many scholars to realise that development might not succeed without state intervention. For example, to Keynse, the state has stronger role to play in economy in order to avoid economic downturn as well as distributional aspect of income.

This sort of reasoning one confined in the US policy of great deal on the aftermath of Second World War. It was also a period of decolonisation in which many nation states found themselves in low per capita income leading to low economic growth and so on. These decolonised countries did also follow the conventional wisdom of the time which suggested that development could be achieved through heavier state intervention leading to higher economic growth. However, the critique of then development paradigm namely women movement argue that this sort of development model completely excludes women perpetually reproduce same patriarchal society and they proposed that unless women are included in development, larger part of humanity could be deprived of economic growth. This paradigm is known as women in development. It was also a period of neoliberalism at full swing started with USA and Britain with so called iron lady market Threacher, whereby state back seated in the realm of economy and market became role focus of economy and politics.

As is indictable above that conception of Human development started with human development report by United Nations, the major theme of writing was coming from Sen. In defining development, Sen has argued that there is no way to see that economic growth reflected development of people. The proponent of Human development is to change human freedom, what Sen Calls as capability and equality, poverty, lack of political freedom like rights etc. make difference in human life. So the conception of Human Security is to enhance the people's capability to reach their potential. A nation state has a role in enlarging people's capability for example economic down turn can make people vulnerable if there is no social safety. Indeed, this was the case in 1994 after economic crises in East-Asian countries. Similarly, acute shortage of natural resources like water or income inequality, level of unemployment may pose the security threat leading to war. That is how Human Security entered in development discourse and current paradigm is marked by this discourse. Further, the proponents of Human Security argue the enhancement of human capability is pre-requisite for peace and safety.

The concept of security in contemporary times underwent a change from the state to ordinary people and from military issues to the everyday security of the average citizen. From the South-Asian perspective, the book "*Reintroducing the Human Security debate in South Asia*" edited by Basrur and Joseph (2007) gives an insight about the region that after attaining independence from colonial occupation, Human Security was the overarching norm. The book mentions Axworthy who argues that Human Security is the broadest conception of security and includes "security against deprivation, an acceptable quality of life, and a guarantee of fundamental human rights". The people in the South Asia believed that freedom would entail them broad based security and opportunities for self-development. However, the immense Human Security deficit encompassing by interstate tensions and the lack of human and financial resources, the region remained in the state of poverty and insecurity. Under such environment, Human Security emerged as an overarching norm and gained popularity in the region of south Asia.

Human Security as a concept is broad based and needs careful examination so as to understand it clearly. In this direction, Ole Hamson in his work "*Madness in the Multitude*" (2002) starts with the security from classical times to the contemporary times. The work starts with the emergence of security concept in classical liberal tradition in which different models of governance were purposed and critically evaluated, one might see in the writings of Hobbes, Locke or Kant. Although Hobbesian notion of security was confined to sovereign state in which the governance would give some amount of security to its people but it leave the ample room for international threat and how could it be avoided. Locke going one step forward from Hobbes by advocating the sovereign state must give individual rights and must secure these rights; he also took that stand by which internal security is given priority. However, Kant argues that international peace could be sustained with balance of power and democracy would appropriate structure of governance. Liberalism, taking in line with human Rights, makes its challenges in the case of international community; for instance, can country 'A' be mute spectator if the country 'B' is having ethnic cleansing or gross violation of human rights. A last century events and incidents are enough to see the twist and turn in democracy and how to extend the human rights issues in the domain of other country has been conceptualized and practised.

There are broadly three different conception of Human Security whereby the thrust is individual whether it is human rights/rule of law's interpretation, or safety of people (humanitarian), or sustainable development. In rule of law conception is that treat to individuals steam from denial of basic human right and floating the due procedure of laws which has often been in present in democratic countries. Slightly different conception of Sustainable development in which the distributive aspect of justice be highlighted and argued that the environmental degradation are caused by the developed and sheer pursuit of market economy; therefore there must be some sort of distributive aspect of justice in order to bring about equality. The conception of Human Security broader in nature that encompasses the all threats people might (or would) have to face, which would be ranging from treats like Global warming to unemployment. Thus, the individuals are centred of reference in moral framework of conception this regard, the strategies and instruments of Human Security are many in international arena. As writes, Ole Hamson states that the inter-communal conflict and civil violence typically are viewed in the term of a set of causal relationship that manifest due to the denial of human rights.

One of the clear cut understandings about the concept and practice of Human Security is detailed in the edited book *'Human Security: From Security to Practice: case studies from north east India and Orissa'* by Acharya et.al (2011). The Human Security and its origin can be traced in 1994 with the United Nations development programme through its human development report. According to report expanded the notion of security primarily based on military balances and included economic ford, health, environmental, personal, community and political security. The second important aspect regarding Human Security is related by the author to Canadian approach "focussing on freedom from fear" and there by adds "a life of dignity" to freedom from want and freedom from fear. The referent object of security is the individual rather than the state. The Human Security greatly varies from region to region and cannot be applied in the same way across all regions. The author tried to look at the nature of threats connected at local national, regional and global levels and made a study to develop a set of Human Security indicators in the states of Orissa and North East India.

The Human Security conception is a broad based term, to understand the term properly; one should have the clear idea relating to the various indicators and references used in the term. In this direction Alexander Kis in "*The Operationalization of the Human Security Concept*" in 2010 provided a helpful tool to apprehend the notion of Human Security. The author pointed out that although there are different experts regarding the meaning of Human Security conception, but there is a general agreement that promotion of Human Security is vital for securing a better life. Also, there are several indexes and lists available and thus, it would be fruitful for establishing a formula related to status of Human Security measurement. While setting up of such measurement, Human Security index would lead towards the operationalization of the conception. The author set up two reasons for the promotion of Human Security index. Firstly, it would lead the researchers to have a better understanding as well as broad vision in understanding the roots of Human Security and its development mechanisms. As it is difficult to set up a Human Security conception by a single unified definition, but by the indexes provided by different academicians, publications and think tanks, it would be easy to assess the realities of Human Security status. Secondly, such an index can shape the human condition by correlations of various references as well as can develop new bases for sustainable development efforts. Thus Human Security index would be a viable tool towards the advancement of the multidimensional parameters of Human Security conception.

In the context of Kashmir, Mahapatra and Shekhawat (2008) in "*Peace and conflict review*" gives an overall picture of underlying problems of Kashmir. The conflict in the valley resulted into the loss of thousands of lives and large-scale displacement from different parts of the valley. It has affected the important sources of livelihood of local people such as tourism, horticulture and handicrafts industries. In addition, it has led to extensive damage to the infrastructure and forests of the region. Besides this, the violence in the valley has affected the development scenario thereby not only discouraging private investment, but also creating obstacles in effective implementation of the developmental policies initiated by public authorities. The state is unable to generate its own resource mechanisms as well as much needed private investment. The author argues that industrial backwardness of the valley can be arrested if extensive capital investment accompanying with modern technology is used. However, how capital investment partly private and partly public would generate without locating nuances in the interaction between agriculture and industry of Kashmir.

Similarly, Ganguly and Bajpai (1994) in "*India and the Crisis in Kashmir*" argue that Kashmir reflects the growing and larger crisis of India's political institutions. The crises mainly arise due to politicization of the civil services and judiciary, the reliance on the military to restore public order, and the overall lack of probity in public life. The problems of Kashmir stem from external as well as internal causes; the Kashmiri quarrel has been assimilated into the larger India-Pakistan quarrel over nation-building and regional order. Moreover, the crisis in Kashmir has serious implications for India as it has been and will continue to be consequential for secularism, democracy, federalism, and nationalism-the four pillars of India's political structure. The author further holds that the pursuit of a more or less purely military solution is a recipe for further political decay. Instead a better paramilitary force under tighter control of civil administration and working within the law will be effective to reduce the crises situation prevailing in the valley of Kashmir.

Assessing the economic condition of Jammu and Kashmir, Ahamad and Yasmin (2012) in the “Impact of turmoil on the handicraft sector Jammu and Kashmir: An economic analysis” argues that the conflict in valley has negatively affected the growth of small and medium industries along with infrastructure which ultimately leads to economic stagnation. In fact one the major industry of valley namely handicraft has been de-accelerated growth state due to the conflict. However, the author argues that growth of handicraft industry can be accelerated if tourism industry is fully promoted. The deteriorating condition of under employment disguised unemployment and the situation of poverty could be altered if the backbone of valley’s economy is fully restored and exploited.

Another work on the economy of Kashmir can be mentioned here in the article by Prakash (2000) titled “*Political Economy of Kashmir since 1947*”. The author argues that agricultural growth and industrial development has been hampered by mismanagement and inability of the state to intervene. A rent seeking and corrupt entrepreneurship relocate resource for personal benefit. The corrupt bureaucracy, political establishment etc. gain by favouring someone over others. The article views that the development or underdevelopment of Kashmir one can observe by Bardhan's theory of propriety classes, i.e. partly development form Kalcaki's idea of intermediate class in underdeveloped nation. Even though Bardhan and Kalaki's idea of classes is severely contested in the context of India (Harris-white), however, this theory is unable to pin down the social processes through which rent seeking emerge and maintained, keeping the peculiar situation of Kashmir. The author further argues that a state as well as centre policy has created a sent seeking governments led to serious erosion of ideal and aspiration of Kashmir people and life them in sheer dis-illusement.

Assessing the conflict situation and its effects in the Kashmir valley, Singh (2011) in "*Conflict in Jammu and Kashmir*" observes that the armed conflict in Jammu and Kashmir have costed the lives of thousands of people and displaced several thousands more. The economic costs too have been devastating. In the last two decades, over 13,500 civilians and 4,500 security men have lost their lives in the state due to the conflict (Annual Report 2008-09, Ministry of Home Affairs, Government of India). The author further argues that since the beginning of the armed conflict, violent incidents have become an everyday occurrence in the state. To tackle the situation special laws like the Armed Forces (Jammu and Kashmir) Special Powers Act, 1990 and the Public Safety Act, 1978 have been promulgated. But these Acts have been associated with allegations of human rights abuses. Also, one of the visible impacts of the current conflict has been the gradual polarisation of state's population along regional lines.

Kazi (2007) in "*Between Democracy and Nation: Gender and Militarisation in Kashmir*" emphasis on the militarisation of Jammu and

Kashmir involving both Indian military forces and militants. The thesis focuses upon the growing influence of military as well as its deep impact on the state and society. It also explains that the most serious price of using the military forces for domestic repression in Kashmir and for military defence is paid by the society and citizens of Kashmir. The author further contends the gender dimensions of the Kashmir conflict vis-à-vis tries to examine the causes responsible for the failure of militarisation in Kashmir so far as the security and justice for Kashmiri citizens is concerned. In the similar manner, Rather (2013) in "*Armed Conflicts in J&K and Its Impact on Society: A Case Study of Kashmir Valley*" examines the causes of the Kashmir conflict and more particularly the resistance provided by the local people to Indian rule who often resorted to violent means with the objective to achieve their right of self-determination. As against this violent struggle, the Indian government deployed around 700,000 military and paramilitary troops to suppress this resistance. All this gave birth to frequent curfews, search operations, crackdowns, arrests, custodial death, torture and disappearances of youth. The author also pointed out that this chaotic environment resulted in shelling, firing as well as the implementation of some controversial laws like AFSPA, PSA and POTA which brings human sufferings at the zenith in the form of human rights violations.

Duschinski (2009) in "*Destiny Effects: Militarization, State Power, and Punitive Containment in Kashmir Valley*" argues that intensive militarization in the valley backed by legal provision yields patterns of license that go beyond the domain of law. These conditions restrict the forms of life and generate disorders and insecurities among the segments of the population. The parameters of militarization in Jammu and Kashmir are established by a series of emergency provisions; most notably the Armed Forces Special Powers Act (AFSPA) of 1990. Under these provisions, the military and paramilitary personnel became responsible for violations including rape, torture, and custodial killing.

The author further argues that since the special acts were designed to protect the country, India has dealt in a military manner as a war in the Kashmir Valley. In this war against the people, Kashmiris seems to be treated as terrorists and enemies of the state excluded from the political arena and exposed to basic human rights because of their collective identity as Kashmiris. In addition, the people of Kashmir are being denied from the 'right to life' due to the presence of such provisions.

The perspectives of national human rights commission in the Kashmir valley has been well documented in another article by Noorani (2000) in "*Kashmir and National Human Rights Commission*". The article views NHRC as a body concerned with the review of the factors including acts of terrorism that inhibit the enjoyment of human rights and recommend appropriate remedial measures". However, In terms of actions by the army and paramilitary forces of the union it can do no more than "seek a report from the central government". It has no investigative powers but perceives keenly an exonerative duty. The NHRC has no time for the custodial killings in the valley. The Kashmir's Human Rights Commission is set up under the same replica of the central law and its remit is confined to state officials only, i e, to violations of human rights by a public servant.

But then the entire raison d'etre of the NHRC is simply to exist - to 'answer' charges of violations of human rights in respect of the insurgency-affected areas and cosmetically for the rest. Its impact on national life has been slender and having not buffing up liberty of human being; NHRC is simply warm puppet of Indian state to reconcile the meta-narrative of international human rights organization, namely USA. However, this meta narrative has never been reflected in the protecting (or enhancing) the live condition of people in the valley.

So far as the human rights abuses in the valley is concerned, Mohiuddin (1997) in Human Rights Violations: A Case Study of Kashmir attempts to analyse the human rights violations in the valley of Kashmir due to the unrestricted arbitrary powers granted to the Indian military and paramilitary forces under special provisions. As a result of this intensive militarisation, thousands of Kashmiri's have lost their lives, prisoned, persecuted, or injured. The article examines these violations in the light of international legal instruments on human rights, Vienna Declaration on Human Rights, the related United Nations resolutions, and the fundamental rights as contained Constitution of India. Also, Habibullah (2008) in "*My Kashmir: Conflict and The Prospects of Enduring Peace*" describes the complex web of issues at the root of the conflict like the ethnicity, national identity, religion, friction between local and national government etc. At the same time, the author deals with the missteps as taken by the government of India but points out that despite the history of betrayals and bloodshed, there exists the greater possibility for lasting peace than ever before. Furthermore, Amin and Khan (2009) in "*Life in conflict: Characteristics of Depression in Kashmir*" argues that the physical, mental and social health are closely interwoven as well as interdependent and over the nearly two decades, the people in the Kashmir valley have been prone to such disorders. The article gives a brief overview about the depression witnessed in the population in Kashmir due to the ongoing armed conflict and its serious consequences on them.

The cost of conflict in the Kashmir valley in terms of human loss as mentioned by Arora and Dhingra (2005) in "*At the Cross Roads: Families in Distress*" focusses on the effects of militancy as well as the displacement of the Kashmiri Pandits. The article also gives a brief outline of how the migration of this community is marked with changes in the pattern of their life style, relationships and family structures. These families have been suffering through financial setbacks and are often frustrated due to the inability to cope up with the family demands which leads towards an increase in the incidence of family disruptions.

Statement of the Problem

There are several debates regarding the conception of human security about its nature, threat perceptions, role of nation-states and so on. It will be here informative to conceptualise as well as critically examine the set notions of human security. In the context of Kashmir, the valley is marred by several factors ranging from gross human rights violations, militarisation, and inability of institutions to deliver justice to stagnation of economy. Under such hardship, it is difficult to see diagnosis of issues in coherent manner. Besides, the bilateral dispute between India and Pakistan over the valley possesses difficult challenges for the people to live secure and flourish life. Here lies the complexity to analyse security not only in terms of violent conflict/ threats but also the human life that provides some level of ground of having hope. Hence, the conception of Human Security allows one to critically analyse these overarching problems in systematic manner which is not done (or less explored) in any existing literature.

Objectives

- To explore the Human Security concept in conflict prone region of India namely Kashmir valley.
- To map the various parameters of Human Security challenges prevailing in the valley of Kashmir.
- To examine the adequately (or inadequacy) of the concept of Human Security in capturing the nuances of the Kashmir valley.
- To analyse the military intervention in Kashmir from different perspective, special attention to be paid of Human Security.

Research Methodology

This research takes Quantitative as well as Qualitative methods to present the case of Kashmir valley as it is having distinct features like claim and counter-claim over its territory. The operationalization of Human Security concept is elaborated through analysis of various studies done so far. An analytical and content analysis method is used to evaluate the changing perspective of policy over a period of time. It includes policies from central as well as state government. In order to analyse changing perspective and its inability to capture the depth of problem, the concept of Human Security is probed. The primary data from various reports like UNDP Report, Economic Survey Report of J and K Government etc is utilized in order to develop the comprehensive understanding about Kashmir. Lastly, the method of focus group is adopted to have glimpse of people perspective on valley problem and potential diagnosis of issues. By doing that, the challenges as well as adequacy (or inadequacy) of Human Security are captured.

CHAPTER – 2

HUMAN SECURITY: EVOLUTION AND CONCEPTUALISATION

"The world will never be secure from war if men and women have no security in their homes and in their jobs"

Security is one of the key issues in international politics. Traditionally, security analysis focussed on the security of states, seeing this security as a function of the levels of threats that states faced from other states in the inter-state system and the manner and effectiveness of state responses to such threats (Rajagopalan 2012). The focus until recently has been what might be characterised as traditional security issues giving primacy to states. The referent object being the state and measures to adopt such policies towards protecting the national interest and values such as sovereignty, territorial integrity, economic and political stability etc. was the earlier idea. The threats at planetary level such as the global warming, international terrorism, dangers of nuclear accidents and breakdown of global monetary system etc. were viewed outside of the control of the state and which largely affect the human kind on this earth. Under this purview, there had been a paradigm shift particularly after the end of cold war in understanding the concept of security i.e., the non-traditional notion of the concept of security emerged and as such the area of human security came into picture.

After the end of cold war, the notion of security has come to embrace the protection of communities and internal violence apart from external attack. It is considered that security at the individual level is related to the security at national and international level. However, the notion of Human security is not confined to any event that occurs within geographical limits. It considers that the individual remains no longer only frightened by the battles and weapons, but also when he is hungry, when his home is threatened, and when his day-to-day life is unsafe (Annan 2001). The apparent distinction from conventional security discourse to human security underlie in protection from not only threats but enabling condition to make individuals life flourished.

The security and safety with the protection and upkeep of individual rights is the prime concern of human security. "Freedom from want" such as poverty, diseases, hunger and "freedom from fear" such as violence, war, conflict etc. is what human security entails. Although, there is no unified definition of human security, but the idea of human security is generally developed from the United Nations Development Program (UNDP) report of 1994. The UNDP in 1994 listed out the few aspects of human security domains such as economic, political, environmental, food, personal, community and health security etc. but the list goes long (UNDP Report 1994). As Bajpai quoted from Mahbub-ul-Haq, that human security is not about states and nations, but also about the individuals and people and argued that the world is entering into a new era of human security in which the entire concept of security will change – and change radically. Under this new conception, security will be compared with the security of individuals, not only with the security of their nations or in a different way, security of people and not just security of territory.

Historical Roots of Human Security

The evolution of human security may be traced much before the use of the term. However, it was UNDP's *Human Development Report 1994* that really made human security a common currency among scholars and practitioners of international affairs (Shinoda 2004). The political use of the phrase "Human Security" dates from the period of Enlightenment, when notions of individual's liberty and self-determination were advanced to counter the dictates of government. Along with the themes of human rights and individual's welfare, the term appeared again in 19th century political discourse with the rise of the nation state, along with the formation of regional alliances – to enforce particular global regimes –the term began to be used to designate mutual strategic security, enforced through diplomatic means or military action. This is the sense in which the term was employed for most of the 20th century (Arie 2000).

During the 20th century, the human security was intended to link the freedom from want and freedom from fear, freedoms that lay at the heart of the United Nations. In June 1945, on the San Francisco Conference, the U.S. Secretary of State reported to his government that: there are two fronts for which the combat of peace has to be struggled. One is security front where victory spells freedom from fear and threat. While the second front is the social and economic victory which means freedom from wants and desire. The victory on both fronts can assure the world is enduring peace and stability (Alkire 2003).

However, at that stage, the phrase ‘freedom from fear’ was proposed to indicate freedom from violence, and the phrase ‘freedom from want’ to designate freedom from poverty. The Report thereby, makes a shift from the narrow concept of national security to the broader concept of human security. It signifies a change from an in-depth stress on territorial security to a much deeper stress on people’s security; and from security through armaments to security through sustainable human development (Shinoda 2004).

The concept of human security can also be traced to the notions of development that took place in the 1960s, 1970s, and 1980s. In the early 70s, the Club of Rome group produced a series of volumes on the “world problematique” which were designed on the idea that there is a complex concern which troubling human beings of all nations: poverty, environment deprivation, loss of faith in institutions, employment insecurity, dehumanisation, rejection of traditional values, and economic disruptions (Bajpai 2000). The Report argued that there is the complex global system influencing the individual’s life and needed a proper attention to overcome these insecurity dilemmas.

In the 1980s, the two other independent commissions contributed towards the transformative thinking on development and security concerning human-centred development. The first was the Independent Commission on International Development Issues chaired by Willy Brandt which, in 1980, issued the so called “North-South report” (Independent Commission on International Development Issues 1980). The report contended that the simplest common interest is that human beings wants to survive, and one might even add has the moral responsibility to survive. This not only advances the questions relating to peace and war, but also how to overcome world hunger, suffering of human beings and alarming inequalities between the living conditions of rich and poor. The crux of the report was the spirit to overcome dangerous tensions and to produce significant and useful results for nations and regions but primarily for human beings in all parts of the globe.

The second commission of the same decade was the Independent Commission on Disarmament and Security Issues (chaired by Olof Palme), focussing on common security. The report gave primacy to military issues and national security but subsequently ushered the need to recognise issues like poverty and deprivation particularly in the developing world due to economic disparities. The Report also noted that Common security requires people to live with honour and peace, that they have plenty to eat and are able to find daily livelihood and live in a world without poverty and scarcity (Bajpai 2000).

After the end of the Cold War, a new thinking in security concerns developed rapidly. The Stockholm Initiative on Global Security and Governance in 1991 issued a call for Common Responsibility and focussed on the threats that stem from failures in development, environmental deprivation, uncontrolled population growth and movement, and lack of progress towards democracy (Government of Sweden 1991). The initiative viewed that concept of global security must be broadened from the traditional focus on the security of states to include the security of people and the security of the planet. These commissions and reports may be viewed as the precursors to human security thinking that in a way helped to shift the focus of security analysis from national and state security to 'security for the people'. However, it was only in 1994 that the term human security appeared in mainstream development circles through the global United Nations Development Program (UNDP) report. The report widened the scope of security from external aggression to include the safety and security of individuals and groups from threats like hunger, disease and political chaos; and protection from sudden and hurtful troubles in our daily lives. The report went on to further identify seven core elements that reflect the basic needs of human security to include economic, food, health, environmental, personal, community and political security (UNDP 1994).

The human security concept is enriched by extending its realm in democratization and civic education in the Costa Rica (1994), by then President Oscar Arias; later on it has been incorporated in the conception whereby demilitarization and disarmament becomes integral ingredient paved the way for increasing human security concerns in many countries (Jolly and Ray 2006). Moreover, after 1994, the concept of human security became a major figure of many governments through their Defence and foreign policies. Since 1995, the leading contributor to the human security dialogue has been the government of Canada, Japan and Norway led the way in institutionalizing human security concerns into their respective foreign policies (Arie 2000).The governments of these countries viewed human security as a tool to provide safety for people from both violent and non-violent threats and also to ensure people's rights, safety and security or even their lives.

In the 1999, the secretary general of UN- Kofi Annan adopted the human security agenda in a personal quest for a new UN mandate labelled under Millennium Declaration to encompass economic progress, social justice, environmental protection, democratization, arms reduction and respect for human rights and the rule of law (Tajbakhsh 2005). It was intended to involve the UN in a more global forum to influence and exert pressure on governments in order to implement more feasible development agenda. The report however laid greater stress on human-centred approach to security but did not provide a formal definition of how this should be articulated or measured (Arie 2000).

In 2001, the Commission on Human Security, chaired by Amartya Sen and the Sadako Ogata was set up to explore the concept of human security and to recommend for policy implication. The report illustrated the connection between well-being and state security and how individuals (or communities) are subjected to threat not only from violent conflict but the obstacles (or insecurity) of achieved well-being (Jolly and Ray 2006). By 2005, attempts were made to clarify human security threats and what the international community should do about them under the theme of two important UN reports. One was the report from the UN High-Level Panel on Threats, Challenges and Change issued in December 2004 and the other was the reform agenda proposed by Kofi Annan in Towards All Freedom (Tajbakhsh 2005). These reports focus the need to readjust to the new realities of the 21st century and to found means of mounting collective responses to new threats like terrorism, the doctrine of preventive interference, and humanitarian interference in the name of human security.

Conceptualisation of Human Security

The concept of human security has been too long narrowly viewed as security of territory from external aggression or as securing the national interest in foreign policy or as global security from the threat of a nuclear holocaust (Human Development Report 1994). It has been related more as state centric than people centred. However, the paradigm of security began to shift with the end of cold war and conflicts began to arise within nation-states rather than between nations. It also led to a feeling of insecurity among people about their physical security and protection of basic liberties in their daily life. The cause of these conflicts were seen to be linked to non-state and non-traditional factors such as internal socio-political conditions, swiftly deteriorating economic conditions, environmental hazards and identity politics (Human development report 1997). It was in this context that the concept of human security has evolved. As such, human security served as an instrument to reduce the human costs of violent conflicts, to counter balance the inequalities of globalisation, to enable governments to address basic human needs and to provide social safety to the poor and marginalized people from severe economic crises (Acharya 2001).

There are various scholars apart from the policy makers who tried to explain the concept of human security under the paradigm of the national security. In due course of time, various theories developed in the study of international relations which deals with the idea of security in various aspects. Theoretically, the notion of human security has been upheld by the liberal school of thought. In international relations, liberalism entails its concern for welfare of the states, along with the individuals as well. It tries to explain the role of the state and non-state actors in promoting security of individual's i.e. human security (Burchill 2001). Also, the post-cold war witnessed severe challenges posed to realism by alternate approaches to security, shifting the focus from the State to the individual and from a military centred concept to a people centred concept of security. The Copenhagen school argued for broadening the security agenda to include the individual as a referent object of security (Tsai 2009). Barry Buzan, Ole weaver and others redefined the concept of global and national security by privileging individual security as the prime concern.

After the end of cold war, the concept of human security appeared on the international scenario and attained a top priority concern at the global level. Respect for sovereignty was shaken by too many examples where states themselves became causes of insecurities, not only failing to achieve their responsibilities toward their subjects but threatening their very existence. Besides, the dangers of global confrontation and major inter-state conflicts decreased with the end of bipolarity. At the same time, new actors appeared on the scene: international organizations, private investment companies, NGOs and non-state entities began to play an active role in international relations. However, this was marked with the emergence of new threats like intra-state conflicts, ethnic confrontations, terrorism, and forced displacement, extreme poverty, HIV/AIDs etc (Tajbakhsh 2005). Human security was a step in that direction to confront with these threats.

The human security owes its origin to the 1994 Human Development Report of the United Nations Development Programme (Acharya 2001). The second important contribution towards human security is the Canadian and Japanian approaches which consider it as the guiding principle of their foreign policies. However, the fundamental goal of the human security agenda is to support and help the vulnerable people and to provide the means to combat with them. So far, various approaches have identified two groups of means or measures: one is preventive and the other is reactive. The Preventive measures denote to all methods that intend to minimize the vulnerability of human beings, while reactive means are assumed to provide human security as secondary means. The use of force is interpreted only as a last alternative (Prezelj 2008).

2.2.1. Human security and the UNDP

The Human Development Report imports to provide a systematic substituent to conventional notion of security and a necessary supplement to human development. It has provided a comprehensive conceptual model of Human Security by shaping the main orientations towards the development of practical values of human freedoms (Kis 2010). The UNDP's conception of human security is based on two principal components i.e, freedom from fear and freedom from want. This kind of security presented protection from chronic threats such as disease, hunger and repression as well as safety from sudden and harmful disruptions of daily existence. It deals with the treatment of security values by two folds: firstly, it makes a number of general statements about values which emphasize the safety, welfare and dignity of individual human beings in their daily lives and secondly it makes aware of the dangers faced at every turn by ordinary men and women (Bajpai 2000).

The UNDP report further argued that for many, security symbolized protection from the threat of disease, poverty, unemployment, offense, social conflict, political repression, and environmental dangers. Hypothetically, human security is a kid who did not die, a disease that did not expand, a work that was not cut, an ethnic tension that did not burst in violence, a rebel who was not silenced. Human security is not a concern with armaments, it is a concern with human life and self-esteem. It is concerned with how people live and breathe in a society, how freely they exercise their several choices, how much access they have to market and social prospects and whether they live in peace or in conflict (UNDP

1994). Besides, there are broadly four fundamental pillars of human security that the report identified:

1. It is universalistic in nature and its applicability.
2. Interdependence of the components of human security.
3. It is preventive in applicability and cost effective.
4. Its locus is individual life rather than state and institutions.

Therefore, the concept of human security intended to extend the orbit of security in broader level whereby, one can apply as well as evaluate the individual's opportunities and threat. The UNDP further distinguishes seven components or specific values of human security: Economic, food, health, environmental, personal, community and political security. Economic security assures every individual a basic or minimum required income. Food security refers to an individual's access to food via his assets, employment, or income. Health security guarantees an individual a minimum protection from various diseases, devastating illnesses and unhealthy lifestyles. Environmental security refers to the protection of people from natural hazards, man- made threats in nature and deterioration of the natural environment. Personal security guarantees an individual from crime and physical violence, whether from the state or from external states or from sub-state actors. Community security refers to the protection of people from loss of traditional relationships and their cultural dignity. Lastly, political security guarantees the freedom to individuals against human rights violations (Henk 2005).

The idea behind the UNDP report was both political and practical. In its political dimension, the UNDP contended that human security offered a ground for the individuals to enable them to live in a society that honours their basic human rights. It also argued that the totalitarian regimes, countries facing political chaos and unrest and those where military expenses are very high; were not likely to be politically secure. The reason behind is that the governments in order to fulfil their political ends and objectives may sometimes use armed forces to supress their own people. Practically, it was expected that an emphasis on human security would make it possible to capture the so-called "peace dividend" and to ensure that the resources devoted to the military through the Cold War were directed towards more productive ends (Krause 2007).

The UNDP also recognised two sorts of threats: localized and global. The localized threats are the threats observed in particular groups, societies and regions of the world. These threats differ from region to region in terms of the level of economic development and geographical location. However, the threats existing beyond the territorial limits of a state i.e, threats within countries are categorised as global threats (Bajpai 2000). To have the clear understanding about the nature of localised threats, the report argues that it is essential to understood them in relation with the seven basic tenants of human security as mentioned below:

- *Threats to Economic Security*: These threats evolve out of deficiency in productivity, large scale unemployment, lack of access to basic facilities and inadequacy of financial security in terms of vulnerability.

- *Threats to Food Security*: These threats emerge due to the failure to provide basic food facilities including insufficient access to assets, work, and assured incomes.

- *Threats to Health Security*: These threats develop due to the spread of transmitted and infectious diseases, diseases of the circulatory system and cancers, deficiency of safe drinking water and lack of access to health care facilities.

- *Threats to Environmental Security*: These threats are the outcome of limited availability of water and water pollution, cutting down of forests, air pollution, desertification, insufficient access to arable land, and natural disasters.

- *Threats to Personal Security*: These threats arise out of human trafficking, drug trafficking, crimes and physical violence particularly to children and women.

- *Threats to Community Security*: These threats are related to the loss of traditional customs, languages and cultures, collapse of the family, ethnic conflict and discernment, and genocide.

- *Threats to Political Security*: these threats stem from suppression on the part of government, massive human rights violations and military operations.

Apart from localized threats, the report further cited some global or transnational threats showing their presence well beyond any given national barriers (Bajpai 2000). These are categorised under six areas:

- the consistent alarming pressure of explosive growth of population on non-renewable resources - in some scholars account is the greatest threat of entire humanity - leads ultimately to global poverty, environmental degradation, and illegal immigration.

- the mounting of inequalities in world income due to greater consumption and production in the developed world have increased the gap between rich and poor, and the poverty, backwardness and environmental degradation in the developing world have led a troublesome environment in those countries.
- the drift of global migrants due to poverty, explosive population growth and the policies of the industrial countries resulted in the increase of refugees and internally displaced persons leading to the ethnic conflicts.
- the environmental degradation resulting in acid rain and global warming, loss of biodiversity and the damage caused to wetlands, coral reefs as well as deforestation.
- the trafficking of drugs along with containing anti-social activities that have developed into a global and multinational relation.
- the appearance of International terrorism as a global phenomenon and its impact on world community

The UNDP Report has led to judge the concept of human security to be all encompassing for practical purposes. It has argued that human security is the idea that views the individual to be at the receiving end of all security concerns, whereby security is understood in terms of freedom from want and freedom from fear (Floyd 2007). The figure 2.1 on the next page entails a broad vision to understand the notion of human security as per the UNDP.

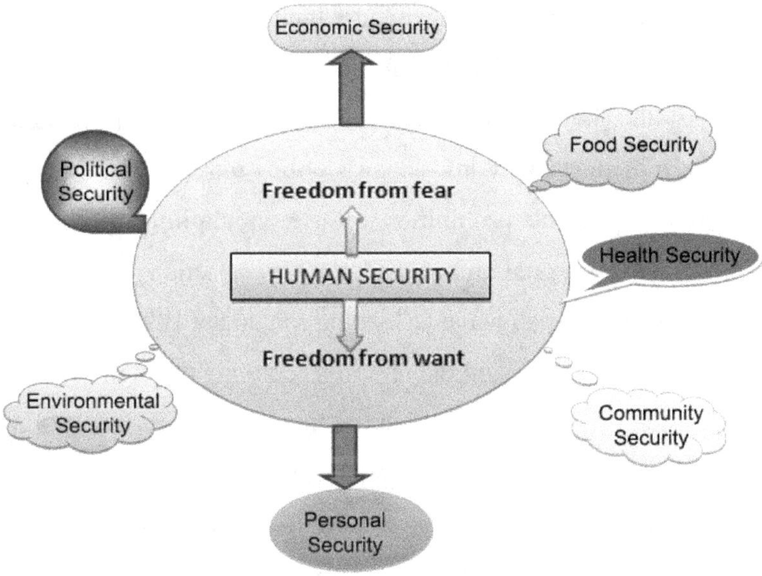

Figure: 2.1. Components of Human Security

Source: UNDP Report 1994

Contribution of Japan to Human Security

Japan has been among the major proponents of the concept of human security. Over more than a decade, Japanese government has continued to uphold the notion of human security besides the successive changes in government regimes from the Liberal Democratic Party (LDP) to the democratic party of Japan (DPJ). To have the clear understanding of this persisting commitment, it is fruitful to analyse the reasons behind the adoption of the policy. In the breakdown of gulf war in1991, the US insisted japan to send self-defence force units to join the US-led coalition against Iraq. Japan, besides being long standing security alliance of US offered only the financial assistance instead of logistic support. This experience ushered its political leadership to redefine Japan's potential role in its global security arrangement which provided a subtle ground for Japan to adopt the notion of human security in its foreign policy.

Another key event that further solidified the incorporation of the concept of human security into Japan's policy making was the signing of the convention on the prohibition of use, stockpiling, production and transfer of anti-personal mines and their destruction in 1997 (Suzuki 2012). In addition, Japan played a leading role in the 1997 Asian economic crises and thereby stabilising the region's economy through Trust Fund by adopting a long-term agenda which the human security was in a position to provide (Tajbakhsh 2005). It led Japanese government to promote the concept of human security with a view to embrace the new economic challenges in Asian region as well as opened a way for Japan to play an autonomous role without being an alliance partner of the US (Minami 2006). As such, human security became a corner stone of Japan's foreign policy and an alternate way to define its overseas development assistance policy. These historical events show that how the human security approaches became a guiding principle for Japan's foreign policy.

Japan has been the vocal champion of the concept of human security. In 1998, Japan's the then Foreign Minister - Keizo Obuchi introduced the concept of human security in his policy speeches (Remacle 2008). Japan holds a wider view of human security encompassing all the menaces that dangers human survival, daily life and dignity, for instance, human rights violations, environmental degradation, poverty, transnational organised crimes, prohibited drugs, illegal immigration, anti-personal landmines and infectious diseases such as AIDS (Paris 2001). The strengthened efforts are needed to confront these threats. Apart from these threats, the government of Japan holds to ensure human freedom by focussing on the individual and non-state actors and aimed at promoting economic development and individual self-reliance. This demands cooperation among the various actors in the international community including governments, international organisations and civil society.

In 1999, the government of Japan supported the creation of United Nations Trust Fund for human security (UNTFHS) and in 2000; Japan initiated a Commission on Human Security (CHS), co-chaired by Sadako Ogata and Amartya Sen (Alkire 2003). However, in 2003, the CHS consequently issued its report entitled as “Human Security Now” (Cornelis 2012). The report of CHS argued that human security aims at protecting the vital core of all human lives so as to enrich human freedom and human fulfilment (Remacle 2008). In addition, the Japanese government created an informal network called “Friends of Human Security” in collaboration with Mexico, along with many UN member countries to propagate a vision of human security approach through non-military means.

The Japanese approach to human security concentrates on “Freedom from Want”, one of the two goals set by the 1994 UNDP report. But, the notion of human security for Japan is a broader concept, with freedom from want being no less critical than freedom from fear and therefore, it was necessary to go beyond thinking of human security solely in terms of protecting human life in conflict situations (Cornelis 2012). However, there is no humanitarian intervention (use of force) on the part of Japanese government in terms of ensuring human security (Takasu 2000). Article 9 of the Japanese Constitution prohibits the use of force to solve disputes, thereby, leading Japan with only forces of self-preservation for international security purposes.

Thus, the Japanese approach is largely related with the basic human needs model and lays stress for the promotion of human needs and human development rather than on protection of people from violent conflicts. In addition, emphasis is imparted for the promotion of peace.

Contribution of Canada to Human Security

In 1990s, Canada is crediting of adopting the notion of human security as a key component of its national security discourse. The foreign minister of Canada Lloyd Axworthy has been pioneer in adopting human security as 'much more than the absence of military threat' including an acceptable quality of life, security against economic deprivation and a guarantee of fundamental human rights (Axworthy1997). He renowned the requisite to restore the Canada's foreign policy by adopting new measures to address the problems of post-cold war period such as menaces of terrorism, increase in drug trafficking and arms smuggling and the condition of children caught in war zones. Further, he called for addressing these problems by shared responsibility through humanitarian interventions (Tajbakhsh 2005). However, the core elements adopted by the Canadian government in terms of human security includes "security of the people" and the identification of UN Charter, the Universal Declaration of Human Rights and the Geneva Conventions (Acharya and Acharya 2001).

The Canadian formulation much focussed on the 1996 UNDPs report which focusses on freedom from fear and thereby calling for the safety for people from both violent and non-violent threats (Axworthy 1999). Quoted in Leaning, M.D., S.M.H. and Arie (2000), the report on human security, the government of Canada entails:

> *"In essence, human security entails safety for people from both violent and non-violent threats. It is a situation or state of being characterized by freedom from pervasive threats to people's rights, their safety, or even their lives. ... From a foreign policy perspective, human security is possibly best understood as a shift in perspective or orientation. It is a substitute way of looking the world, taking people as its reference point, rather than focusing entirely on the security of territory or governments. Like other security concepts – national security, economic security, food security – it is about protection. Human security involves taking preventive measures to reduce vulnerability and minimize risk, and taking remedial action where prevention flops."*

The Canadian government included the conception of human security in its foreign policy agenda as an attempt to play a greater role in international scenario and also to reduce threats to its own security as being a middle power. The human security agenda undertaken by Canada led it to play a leading role in the campaign of banning against anti-personal landmines. The other important outcome of this policy is the signing of Ottawa treaty in December 1997 (Remacle 2008). The treaty was signed by 122 countries on the Prohibition, Practice, Storing, Production and Transfer of Anti-Personnel Landmines and their Destruction. Further, the agenda had led the Canadian government to adopt the policy of peace building to address complex challenges of societies in conflict and to manage conflict without violence by focusing on the political and socio-economic context of internal conflict (Acharya and Acharya 2001). According to Tajbakhsh (2005), the human security policy of Canada is prioritised in five directions which are inter-related:

- The safety of citizens from threats like drug trafficking, arms smuggling, illegal immigration etc.

- Protection of citizens through legal norms, demilitarization and operationalization of human rights.
- Preservation of conflict to reduce tensions whether local or international without using the violence.
- Deepening the governance, making it accountable and promoting freedom of expression to its citizens.
- Supporting peace in international relations.

Besides, the government of Canada in collaboration with the Norwegian government established a Human Security Partnership under a joint Canada-Norway initiative in 1999. This understanding of human security was jointly adopted by other middle powers under Human Security Network (Remacle 2008). This network was launched primarily to discuss the international campaign against the banning of landmines, but later its meetings emerged to cover a wide range of issues. By 2004, 12 countries joined the network including Canada, Greece, Austria, Chile, Ireland, Jordan, Mali, the Netherlands, Switzerland, Slovenia and Thailand while South Africa played a part as an observer (Henk 2005).

The network served as a platform to strengthen the political process and to encourage peace and development by curtailing the conflicts within the countries of network. In addition, it also led to the inculcation of NGOs with a view as to create a more humane world (Paris 2001). It recognised a nine point agenda for human security including elimination of landmines, establishment of the International Criminal Court, human rights education and human rights law, proliferation of small arms, struggle against international crime, fight against disease like HIV and AIDS, child labour, children and women in armed conflict and cooperation of the developed world (Henk 2005).

In summary, the Japanese approach to the notion of human security recognises critical security threats but provides a tactics to cope these threats by intending to foster the economic growth and prosperity as well as to reduce or prevent conflict. On the other hand, the Canadian approach mainly focuses and attaches priority to the protection of people in times of war and complex emergencies.

Approaches of Human Security

Human security is an all-encompassing concept which includes within its ambit a wide range of neologisms ranging from security of people to security of community; and from cooperative to comprehensive security. Broadly, there are three schools of thought/approaches to understand the concept of human security: Rights-Based Approach, Humanitarian Approach and Sustainable Human Development Approach.

Rights Based Approach

The rights based approach also called as rule of law/ human rights approach have its roots in liberalism focussing on individuals basic rights and the responsibility of state towards granting and protecting these rights. Its focus of concern is the promotion of rule of law, democratic system of governance as well as securing of fundamental human rights (Donnelly 1993). It considers international institutions as instruments in providing new human rights rules and for bringing about a convergence in different national values and practices (Hampson 2002).

The approach is largely based on wider definition of human security, aimed to strengthen value based lawful structures at both national and international levels. It also lays emphasis about the denial of minority rights which under certain circumstances could pose a threat to human security (Kymlicka 2000). The threats arise out of weak judicial and political system, hence the need to strengthen international norms and provisions.

At global level, the belief of respecting the human rights and minority rights were being considered necessary for the promotion of peace and security more particularly with the promotion of nation-state. The treaty of Westphalia not only came up with the principle of sovereignty but also with the guaranteed rights of religious minorities (Krasner 1999). Also, the establishment of special minorities committee in the League of Nations was a step forward in this direction. With the end of Second World War, the United Nations came into being primarily to maintain peace and cooperation in the world. Article 1 of the UN charter "reaffirms faith in the equal rights of men and women and of nations, in fundamental human rights and in the dignity of a person" (Hampson 2002). Further, the UN declaration of human rights in 1948 identified certain rights including personnel rights, legal rights, subsistence rights, political rights, economic rights etc.

In addition, certain UN Conventions and Covenants have been added to attach great importance and meaning to these rights including the international convention on the elimination of all forms of racial discrimination(1965), the international covenant on civil and political rights (1966), the international covenant on economic, social and cultural rights (1966), the convention on the elimination of all forms of discrimination against women (1979) and the convention on the rights of the child (Lauren 1998).

The rights based approach apart from focussing on international institutions aimed to provide different strategies for the promotion of human security. The first instrument is to deny access to domestic groups from foreign commodities, services and capital which would generate interest among the domestic elites and led them to shift their attention in favour of greater protection of human rights (Doxey 1996). Another way is the establishment of international bodies to monitor the human rights violations and also containing information about the people whose rights have been denied or violated should be made available publically. Finally, the establishment of the regional or international courts having the authority to judge and enforce human rights which, ultimately leads to the promotion of democracy (Moravcik 1995).

Humanitarian Approach

This approach is often described as safety of peoples approach and provides a humanitarian basis to the notion of security. It primarily focuses on securing the moral and legal rights of citizens in times of violent conflicts and war. It considers war the greatest threat to human security (Hampson 2002), as it leads to large human rights violations, atrocities and huge loss of lives of citizens apart from combatants. The approach viewed that international interventions are necessary for securing the safety of people. Also, international law needs to be strengthened for the purpose of reducing war-crimes as well as abandoning of all armaments of warfare that are harmful to civilians. That is the reason many intergovernmental and non-governmental organisations have their origin in wars. In addition, the approach also laid emphasis to provide assistance and relief to the people harmed or injured during wars or residing in conflict situations.

After the end of First World War, the League of Nations was created for the purpose of avoidance of war and subsequently the UN set up this as its ultimate objective and purpose. In terms of helping the victims of armed conflict and war, many UN agencies were established including World Health Organisation (WHO), United Nations Childrens Fund (UNICEF), United Nations Relief and Works Agency (UNRWA), and United Nations Development Programme (UNDP). (Moore: 1996). Also some non-governmental organisations have come up to play a pivotal role safeguarding the interests of victims as well as advancing the human security needs. Some of them are Red Cross, Catholic Relief Services, Doctors without Borders and the World Vision (Anderson 1996).

The humanitarian approach apart from concentrating upon avoidance of war, strengthening of international law and relief assistance also focuses on the peace building and conflict prevention operations for maintaining peace and reducing human rights violations. It further emphasis upon the need of redefining the role of military and police forces in the maintenance of law and order, rebuilding small or medium firms, reformation of economic policy and the needs of vulnerable groups of society particularly women and children (Boutros-Ghali 1992).

Sustainable Human Development Approach

The approach is largely based on the United Nations development programme (1994) and includes within its ambit a wide variety of security of security concerns such as food, health, economic, environment, personal, community and political security. It also focuses with the impact of security in the individual in concern with the world economic disparities, globalisation, environmental degradation, illegal immigration, uncontrolled population growth, drug trafficking, arms smuggling, global terrorism and diseases like AIDS etc. However, the conception traces its genesis from 1960s by the adoption of Covenant on Social, Economic and

Cultural rights which guaranteed the right to the security of an individual and also the social security (Hampson 2002).

In addition, the Brandt Commission in 1980s argued about security not only in military terms but also on the global cooperation and sustainable development and prosperity based on equally shared resources (Independent commission on international development 1980). Also in 1987, world commission on environment and development gave an expression of sustainable development and argued about the maintenance and promotion of peace, development as well as environmental concerns (World commission on environment and development 1987). The report mainly focussed on the environment and viewed that environmental degradation leads to political tension and military conflicts.

In short, the approach mainly deals with the instruments of human development more particularly at global level by visiting through socio-economic inequalities. Also, the civil society and its organisation have to play an important role in the advancement and execution of human security concerns apart from international institutions. Thus, these three approaches generate a sense of understanding in the evolution and promotion of the notion of human security at international level.

Addressing the Challenges to the Concept of Human Security

In the post-cold war period, the concept of human security evolved as a serious challenge to traditional notion of security in order to address a series of new security threats like global poverty, intra-state clashes, illegal immigration and scarcity of resources. It challenged the state centric traditional notions of security by proposing that individual people, not nations should be the focus of future security approaches (Cornelis 2012).

Since the inception of the phrase ‘human security’, its proponents have struggled to put a theoretical edge to the concept. However, most of the exponents of the concept argued that the ultimate threats to security arise from internal conflicts as well from state itself rather than from external opponents. It does not led to suppose that human security is certainly in conflict with state security; the state remains the chief source of security in ideal situations, but the over-emphasis upon state security can be impartial to fulfil the human welfare needs. So, the conventional notions of state security are compulsory but not adequate conditions of human welfare.

The concept of human security has attained great attention particularly after 1990s by shifting the focus from traditional security concerns to the individual. The proponents of the broader approach based on UNDPs report of 1994 argued that any conception of security which neglects the threats and security concerns as mentioned in the report is conceptually, empirically and ethically inadequate (Newman 2001). The critics of the broad vision argued that human security is nothing more than a shopping list; it involves slapping the label of human security on a wide range of issues that have no essential link. At a certain point, human security becomes a loose synonym for ‘bad things that can happen’ which results in loosing all utility to policymakers and incidentally to analysts (Krause 2007). It has also been challenged on a number of conceptual and methodological grounds of being so inclusive that the term becomes largely meaningless as an analytical tool. If every human dilemma is a security issue, so the argument goes, how will it be possible to recognise, classify, and prioritize what actually must be protected? (Henk 2005).

Further, the term human security itself is ambiguous, slippery in design and its appeal lies in its very vagueness.

The criticism is also posed in considering the definition of security. Conceptually, the difficulty arises out of endless debate about the threat spectrum as presented by the supporters of human security. The problem here lies is that who is to provide human security? The individuals themselves are in no position to provide security of their own, then how should they claim their own right to survival (Floyd 2007). Thus, logically the notion of human security can only be protected as well as guaranteed by the larger entity like state or any international institution.

The concept of human security is criticised in terms of giving different prioritization to different threats as freedom from fear is sometimes mentioned as a key element of human security; at other instance, acceptable quality of life is considered inevitable. This prioritization makes it difficult for the academicians and policy makers to implement and practice the term properly with regard to threat perceptions. Further, the concept is challenged on the multidimensional nature as it denotes all aspects of human life, multiple of security threats and a wide ranging variety of means. Such an all compassing concept essentially lacks logic as well as focuses and becomes difficult to implement it in practice (Prezelj 2008).

Moreover, the objection is raised in the exertion of drawing clear and rational connections between and within the component parts (Henk 2005) as well as in establishing a causal relationship between political and economic issues or between the security of individual and the preservation of international peace which seems unjustified. The critics have also challenged the human security policy as adopted by some nations like Canada and Japan. They argued that the Canada adopted the conception of human security in its foreign policy to enhance their role as well as to draw attention of the world towards middle power influence and status in the international scenario. On the other side, it is stated by the critics that the Japanese government in order to promote its own economic interests adopted the notion of human security in one of its policies (Tajbakhsh 2005).

The critics of this conception further challenged it on the grounds for shifting the emphasis of security from the collective will of people to a wide range of individual and collective political, economic, and cultural rights as defined by international bodies or non-state actors like NGOs. This has diminished the very nature of nation-states and roles and responsibilities they have to play (Gasper 2008). In addition to all this criticism, many scholars and experts in international relations have raised doubts about the concept of human security. Quoting Jolly and Ray (2005), there are at least five points which remained in discussion around this uncertainty as:

- Human security simply comprises renaming many problems that have already been renowned in other frameworks and that already have perfectly good names. What has extended by relating them under a new title?
- Human security lacks definite limits and visualises everything as threat to security which sorts the nature of the term as ambiguous and makes it difficult to implement.
- Human security while enlarging its focus to broader aspects like global terrorism, diseases and climatic change makes it difficult

for the international institutions to make decisions or taking action on the threats identified.

- Human security jeopardies the military engaging in concerns which would be actually handled through non-military means.
- Human security though raising high hopes with the UN is expecting to play a role beyond its capacity which it cannot accomplish.

Apart from all the criticisms and challenges posed by different scholars, it is generally argued that that the referent object of security analysis should be the individual, but the disagreement is raised only in terms of the threat perceptions and the means by which they should be combated. Thus, human security seems to appear as an endless debate between its advocates and opponents and even among its supporters, who have not agreed on a single unified definition (Acharya, Singhdeo and Rajaretnam 2011).

CHAPTER – 3
UNDERSTANDING THE CONFLICT IN KASHMIR VALLEY

Since the inception of India from British colony as an independent state, one of the major region namely Kashmir valley has been conflict prone area within national boundary as well as at the international level. The conflict in the valley of Kashmir has its roots in the sub-continent's partition into India and Pakistan. Both the states desired to merge the valley to their own sides for broadening their political dimensions. Pakistan believed that Kashmir had to be part of it as majority of the state's population constituted of Muslims. On the other hand, the leaders of Indian National Congress wished Kashmir to make a choice in India's favour as its claim of secular state would be enhanced. Conventionally, Kashmir has been viewed from bringing peace through military intervention.

Historical Roots of the Conflict

The state of Jammu and Kashmir comprises of three regions including the Kashmir valley, Jammu and Ladakh. The chapter will mainly focus on one region, i.e., Kashmir valley. The origin of the conflict in Kashmir valley can be traced back to the British legacy in Indian history.

In 1846, the British government concluded the Anglo-Sikh war under the Treaty of Amritsar by transferring the state of Kashmir to Maharaja Gulab Singh in lieu of seventy-five lack Nanukshahee rupees. By this treaty, Jammu and Kashmir became one of the princely states under the colonial rule till 1947 (Marwah 2007).

During the reign of Dogra rulers, the majority of Kashmiri's as being Muslims witnessed the loss of control over best agricultural lands and lagged behind in wealth and education. It led the Muslim majority of the valley to rise in revolt against the feudal rule of Maharaja for political, economic, cultural, and religious rights (Wani and Suwita 2013). Under such circumstances, the people of Kashmir started the first organised movement under the leadership of Sheikh Mohammad Abdullah in 1931, known as Jammu and Kashmir Muslim Conference. During its initial period, the movement focussed on granting of economic, political and other rights as deprived by the Dogra rulers. Besides, the movement make demands on grounds of stopping the discrimination of the Kashmiri people in the recruitment policies of the state (Ganguly 2003).

In 1933, the Dogra rule became unpopular. The British government often articulated concerns over the inequalities of land ownership between the subjects- Hindus and Muslims, lack of quality in Muslim education and a minimal share in employment concerns. It also appointed the Glancy commission in 1932 to investigate Dogra tyranny against the Muslims and demanded reformation within the state of Jammu and Kashmir. This forced the then Maharaja of Kashmir Hari Singh to establish an elected assembly identified as the Praja Sabha. The Sabha consisted of 37 members out of whom 35 were to be selected by communal constituencies and to guide the maharaja on social and economic policy.

However, its opinions and suggestions were not obligatory (ICG Report 2003).

In 1938, the All Jammu and Kashmir Muslim Conference changed its outlook and to make it broad based converted into Jammu and Kashmir National Conference. At the same time, its leaders came in contact with many other political leaders and parties in the Indian sub-continent resulting in widening the vision of the political leaders of Kashmir. It was in this context that the leaders of Jammu and Kashmir National Conference adopted Quit Kashmir Movement with a view to put an end to the autocratic rule of Dogra's in the valley of Kashmir (Wani and Suwita 2013). On account of launching the movement against the Dogra rule, many NC leaders including Sheikh Mohammad Abdullah were arrested.

The genesis of the Kashmir conflict can be traced back to the India's partition in 1947 on the basis of Indian Independence Act which divided the British India into two independent dominions of India and Pakistan. The Act also legalised the princely states to choose independence or accession to domains of India or Pakistan (Heymann 2003), based on geographical contiguity and demographic structure of the population. The British government assigned the job of princely states to decide their fate with the Lord Mountbatten-the last governor general of India. With the exception of few states like Hyderabad, Junagarh and Jammu and Kashmir, all other states integrated with the Indian union peacefully through negotiations before 15 August 1947 (Adhikari and Kamle 2010). Junagarh, with a Muslim ruler but with a majority Hindu population wished to accede with Pakistan. However, India forcibly occupied it on the belief that majority of its population was predominantly Hindu (Qadir 2009). After the occupation of the territory of Junagarh by Indian government the plebiscite was conducted through which it became the part of India.

The state of Hyderabad wished to remain independent but the ruler of the state (Nizam) as being Muslim decided to accede with Pakistan and had already signed a Standstill Agreement with Pakistan. The people of Hyderabad as being predominantly Hindus revolted against the Nizam on the part of his own desire of signing the agreement with Pakistan. However, the Nizam's police called Rajakaras started crushing the rebellion and dealt with them in a very ruthless way. As a result, India intervened in the matter and assisted its limited military action. The conflict between Indian army and Rajakaras lasted for 48 hours and ultimately, the Nizam's police surrendered themselves which led to the accession of Hyderabad with the Indian union in November 1948 (Adhikari and Kamle 2010).

So far as the state of Jammu and Kashmir is concerned, its the then ruler Maharajah Hari Singh delayed his decision, whether to join India or Pakistan till the Pakistan tribesmen arrived in valley (Qadir 2009). The rationale behind this was that majority of population was Muslims (78%), while the ruler was Hindu. His personal perception was to remain independent and not to join either of the two domains. While, choosing the option of independence was not desirable at that time as the state was not self-sufficient in terms of political or economic conditions to survive independently without joining either of the two domains India or Pakistan (Marwah 2007). However, Mountbatten later announced that the third option, i.e., independence was only a theoretical option. So, there were only two options either to merge with India or Pakistan.

The Jammu and Kashmir state enjoyed one of the distinctive positions among the Princely states in terms of its geographical embededness and socio-psychological mind-set of its people. As per its geographical position is concerned, Kashmir was the only princely state which could have joined either of the two dominions. However, in positions of "the nature of connectivity, the states proximity with areas that constituted Pakistan was much greater, compared to India" (Baba 2012). Ideologically both India and Pakistan viewed Kashmir vital for broadening their political dimensions. Pakistan believed that Kashmir had to be part of it, as majority of the state's population constituted of Muslims. In addition, Mohammad Ali Jinnah- the founder of Pakistan admitted that 'k' in Pakistan stands for Kashmir and without its presence Pakistan would be incomplete. On the other hand, the leaders of Indian National Congress wished Kashmir to make a choice in India's favour as its claim of secular state would be enhanced.

The ruler of Jammu and Kashmir was actually confronted with a dilemma regarding to merge either of the two dominions on two grounds. Firstly, he was conscious of the fact that by acceding to Pakistan, he would be never allowed to continue as a ruler as being a Hindu in a Muslim majority state. Secondly, by acceding to India, he had also problems as he had no faith on the leaders of Indian National Congress because they have always criticised and opposed the rule of princes. Apart from it, these leaders had strongly supported the popular Kashmiri leader-Sheikh Mohammed Abdullah whom the maharaja had thrown in jail for launching the Quit Kashmir Movement (ICG Report 2003).

Under such circumstances, the Maharaja of Kashmir entered into an agreement with Pakistan known as Standstill Agreement for the maintenance of necessary commodities including the food, salt, oil and petrol as well as for the promotion of essential services including railway, telecommunication and banking. However, Pakistan had no faith on Maharaja and did not give him time to consolidate his position but, instead exerted pressure on him to accede to Pakistan (Marwah 2007). The Pakistan also stopped the essential supply of commodities and thereby notwithstanding the terms and conditions of the agreement. In addition, the armed Pathan tribesmen from Pakistan led infiltration into the valley, captured various towns and massacred large number of civilians (Hussain 2009). To cope up with these tribesmen, the Maharaja requested the India for military assistance and help.

The Indian government refused to help and instead taking advantage of the situation agreed to provide the support only with one condition that the state should sign 'Instrument of Accession' with India (Baba 2012). Under such situation, the maharaja of Kashmir signed the instrument of accession with India on 26 October, 1947. It is to be pointed out that, "the accession with India was not voluntary rather imposed on the Maharaja under an extra-ordinary political situation following the tribal Muslim invasion". At the same time, the Governor General of India-Lord Mountbatten on 27th October 1947 declared that "as soon as the law and order have been restored in Kashmir and her soil cleared of the invaders the question of accession would be settled through a reference to the people" (Adhikari and Kamle 2010). Besides, a meeting was held between Lord Mountbatten and Mohammed Ali Jinnah on 1 November, 1947. After the meeting, Mountbatten said that, "The accession was not a bonafide one since it rested on fraud and violence and should never be accepted by

Pakistan. Accession was the end of a long intrigue and that it had been brought about by violence" (Tabassum 2012).

Immediately, after the accession, the Indian forces preceded into the valley and driven out the invaders. However, the accession could not yield fruitful results and practically divided the state into two parts. The smaller portion remained under the control of Pakistan whereas, the larger portion came under the control of India. While looking in the legal perspectives of the accession, it can be viewed that the major factor behind the possession of the respective areas was the physical force.[1] The map 3.1 on the next page demonstrates the entire geographical area of Jammu and Kashmir at the time of Indian partition in 1947.

[1] After the 1962 war between India and China, the state of Jammu and Kashmir got divided into three parts and under the authority of three distinct sovereign nations. In terms of the total area, nearly 35.15% is presently under the control of Pakistan, 2.29% had been gifted by Pakistan to china, 16.92% had been occupied by china and the remaining 45.64% is with India.

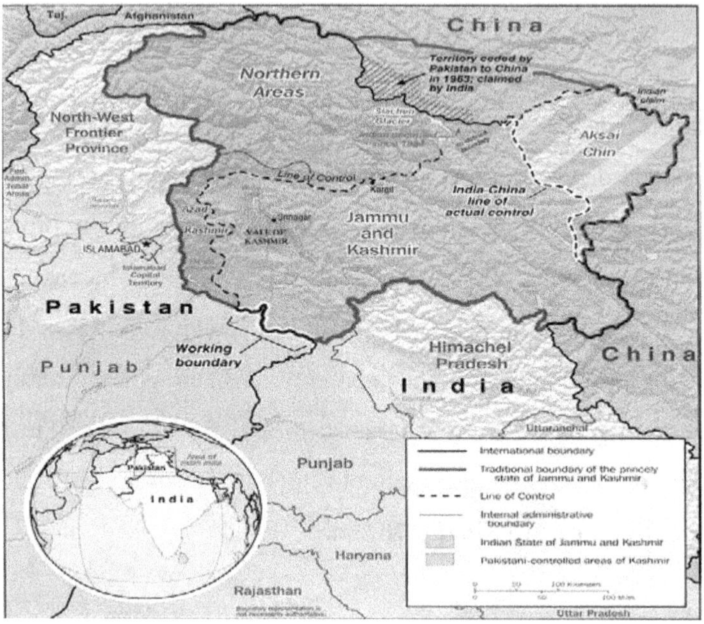

Map 3.1: Map of Jammu and Kashmir

Source: Washington: Central Intelligence Agency, 2002

On behalf of the aggression launched by the Pakistan's tribesmen, the Indian government approached the United Nations on January 1, 1948 to file a complaint under Article 35 of chapter VI of the UN charter[2]. After several days, Pakistan also filed a counter-complaint against India of being accused of aggression in other princely states like Hyderabad and Junagarh. Also, Jawaharlal Nehru the then Prime Minister of India accepted the resolution passed by the UN i.e, plebiscite in favour of Kashmiri people and promised to abide by it but later violated. It had been widely debated that this move of India had internationalised the issue and thus making the 'Kashmir dispute' as an international dispute.

[2] Article VI of the UN Charter maintains the continuance of any state which is likely to endanger the maintenance of international peace and security.

The signing of the Instrument of Accession with India by the maharaja of Kashmir-Hari Singh resulted in a hostile environment between India and Pakistan leading to three major wars between them on Kashmir issue, apart from several border skirmishes. The first Indo-Pak war erupted in early 1948 in the areas around Punch, Kargil and the Ladakh. The Indian forces intended to regain some of the ground which had been under the Pashtun tribesmen. But Pakistan's regular army joined with the tribesmen to counter the Indian army directly (Marwah 2007). India referred the issue before the United Nations and demanded the international intervention of the matter (Hussain 2009). However, it took almost a year for the mediation process to put an end to this hostile environment between India and Pakistan (Ganguly 1995). The United Nations ultimately passed a resolution on 1 January, 1949, under which both countries agreed to sign a cease-fire agreement and thereby making an end to the first war over Kashmir (Hussain 2009).

Though the first war was much lengthy of all the three wars fought between India and Pakistan on Kashmir dispute, but, instead, was the least expensive in terms of both material and human losses. The rationale behind being is that, it was basically a land war, fought in mountainous areas, use of limited arsenals (tanks) and no use of navy forces (Ganguly 1995). The war also did not cost the civilian killings, however, it is estimated that military causalities from both Indian and Pakistani side was around 1500. Apart from it, approximately 5000 square miles or one-third of Kashmir came under the control of Pakistan's territory (Lamb 1996). The war of 1947 was an apparent achievement for Pakistan as it acquired the portion of Kashmir that it still controls today called Free Kashmir or Azad Kashmir (James and Zdamar 2005).

The second war between India and Pakistan took place on 31August, 1965, when Pakistan launched another armed operation against the Indian part of the State (Hussain 2009). It was an attempt from the side of Pakistan to occupy the remaining part of Jammu and Kashmir State by force (Bamzai 1994). The war launched by Pakistan was coded under 'Operation Gibraltar' with the view to serve two purposes: a) to pursue the sympathy of Kashmiri people, supported a massive infiltration of Pakistani rangers into the Valley, thereby, initiating them a spirit to rise against India and b) the larger scale military conquest along the whole Cease Fire Line (Adhikari and Kamle 2010). It continued over more than a month and concluded by a UN sponsored ceasefire (Singh 2011). Though, the ceasefire declared by UN on 6, September was only accepted by Pakistan.

The China threatened India by giving an ultimatum that if India did not put an end to the war against Pakistan and vacates the occupied areas; it would, then face the consequence (Gupta 1970). However, it was only with the efforts of Soviet Union that the heads of two countries i.e, Lal Bahadur Shastri and Mohammad Ayub Khan met in Tashkent in January 1966 to sign the declaration known as Tashkent declaration (Singh 2011). The purpose of the declaration was to solve the disputes between India and Pakistan through peaceful means and removing of troops from both sides to the previous Ceasefire Line. Moreover, the declaration was aimed for a peaceful resolution to the Kashmir dispute but it lacks the mechanism by means of which any resolution of the dispute could be drawn. Hence, the Tashkent agreement concluded without any meaningful gains.

The third war on Kashmir known as Kargil war was another attempt made by the Pakistan to settle the Kashmir dispute by means of military solution. The war started on 26, May and continued until 11 July, 1999. The Pakistani forces occupied the terrain in the Dras-Kargil sector which the Indians used to vacate every winter. The logic behind was to fulfil two purposes. In the political dimension, the operation was undertaken for the purpose of seeking a fair and permanent solution to Kashmir issue. While, in military terms, it was operationalized to generate a military threat and compel India to the negotiating table from a position of weakness (Qadir 2009). The Kargil war was the biggest military operation in which Indian forces suffered heavy losses. Also, the diplomatic relations between the two countries were shut during the war discourse.

The Kargil war of 1999 also concluded in the same manner i.e, without yielding fruitful results and without fulfilling Pakistan's desire about the Kashmir dispute. Practically, the war ended due to the pressure exerted by US and the G-8 countries over Pakistan. Under such circumstances, Pakistan unambiguously accepted the ceasefire as declared by India on 11 July, 1999. By following the guidelines of US and the countries of G-8, Pakistan withdraws all its troops from the region on 16 July 1999 (Adhikari and Kamle 2010).

Kashmir in the United Nations

With the end of Second World War, the United Nations came into being with the purpose of preventing the world from the occurrence of next world war. In addition, it aims to resolve the conflicts and to maintain peace and cooperation among the nations in the world. The issue of Kashmir has been taken to UN for the purpose to resolve it peacefully. The dispute over Kashmir is the oldest unsettled international conflict in the contemporary world. Also, the international community with the exception of India recognises Kashmir as a disputed territory and so does the UN. The people in the valley of Kashmir are struggling for the right of self-determination and to seek basic human rights through the implementation of UN resolutions on the Kashmir issue. Therefore, till the date, Kashmir remains among the fundamental disputes of the United Nations agenda.

On 1, January, 1948, India took the issue of Kashmir to the United Nations and filed a complaint of aggression against Pakistan under article 35 of the UN charter. The compliant contained an assessment of Kashmir's situation from September 1947 including Kashmir's request to India for assistance of tribal invasion. It further provides specific details about the Pakistan's assistance to the tribesmen such as: a) providing arms and ammunitions to them (b) allowing them to enter the valley (c) using the territory of Pakistan as a base for the operation (d) inclusion of Pakistan citizens with them (e) supporting them in terms of transportation and military equipment's; and (f) providing the training and guiding them (Das 1950). However Pakistan denied all such claims and filed a counter compliant against India on 15 January, 1948. The claims on the part of Pakistan included the accession of Kashmir by fraud and violence, forcible and unlawful occupation of the territories of Hyderabad and Junagarh; and genocide against the Muslims. In addition, the object of various acts of aggression against Pakistan is the destruction of the state of Pakistan (United Nations Official Document Report 1948).

Both India and Pakistan agreed on a point that a plebiscite should be held in Kashmir to solve the Kashmir issue. Pakistan held that a free plebiscite can be conducted only when the people who have been cleared or forced to leave the state since august, 1947 are restored to their homes.

The India on the other hand contended that a number of Indian troops should remain in the valley till the plebiscite is conducted for the preservation of peaceful conditions (Ibid). The United Nations Security Council after hearing both the parties passed two resolutions. The first resolution of UN was passed on 17 January, 1948 asking both the countries not to intensify the situation but to take every effort in order to improve it. The second resolution was passed on 20 January, 1948, which led to the creation of a mediatory commission came to be known as United Nations commission on India and Pakistan (UNCIP). The resolution of UNCIP was drafted on February, 1948 based on the consensus of opinion among the majority members of the Security Council. The commission was jointly worked out by Belgium and Canada calling out for speedy redressel of violence and ending of hostilities, removal of all armed forces from the state, return of all native citizens who had left the state and the conduction of plebiscite under the UN supervision as soon as possible. The Pakistan accepted the resolution while India wanted the postponement of the debate (Hilali 2007).

The Pakistan got the support of British delegation and others; and it is noteworthy to mention that Dr. Tsiang of China remarked that the key problem of Kashmir lies in the plebiscite. The only way to avoid violence and use of force would be removed only by conducting a free and impartial plebiscite and for deciding the question of accession of Kashmir to India or Pakistan. Though, UN Security Council took almost three months to take a concrete action in this direction and finally adopted a resolution on 21 April, 1948, based on three principles i.e, withdrawal of troops and armed forces from the state, conduction of plebiscite and formation of interim government in Kashmir (Breacher 1953). However, it became clear that the Security Council was not considering India's complaint against Pakistan's aggression on Kashmir but was more interested in locating of India's action in Kashmir (Das 1950).

Meanwhile, the United Nations commission for both the countries arrived on July, 1948. After investigating and assessing the situation in Kashmir by consulting both the parties, the UNCIP adopted a resolution on 13 august, 1948 (United Nations Official Document Report 1948). The resolution stated that the accession of the state of Jammu and Kashmir either to India or Pakistan will be decided through the democratic method of a free and impartial plebiscite. However, the plebiscite will be conducted only when the commission finds that the agreements i.e, ceasefire and withdrawal of troops are being properly followed by both the countries. After that, the United Nations in agreement with the commission shall nominate a plebiscite administrator having the quality of high international understanding and commanding general confidence (Ibid). However, it was only in December 1948, that both India and Pakistan accepted the commission's proposals. Accordingly, a military advisor was appointed by the commission to assist in the implementation of the UN resolution of 13 august, 1948. In addition, the commission also received a group of military advisors from the countries of Belgium, Canada, Mexico and the United States (Hilali 2007).

On January 5, 1947, the UNCIP recommended a plebiscite as the only way of solving the Kashmir dispute prior to the agreements of ceasefire and withdrawal of troops from both the sides of the state of J and K. Again the disagreements between the governments of India and Pakistan appeared on the issues of disposal of Azad Kashmir forces and removal of regular forces from the state. On March 21, 1949, the secretary general of UN appointed W. Nimitz as the plebiscite administrator for the purpose of submitting the disagreements on the part of both the countries over the implementation of the resolutions of the august 13, 1948 and January 5, 1949. Pakistan agreed while the India refused and the whole matter was referred back to the United Nations (Ibid). Though, the UNCIP failed to bring both the parties on the agreement but it recommended that the demilitarization must be treated as the whole; and also the five member commission should be replaced by single individual (United Nations Official Document Report 1948).

Mediation efforts by General McNaughton

The United Nations Security Council adopted a resolution on December 17, 1949, by appointing Andrew McNaughton, the then president of the council as a mediator between India and Pakistan. He had to meet informally with both the parties vis-a-vis finding the possibility of settling the Kashmir problem and to report back any progress to the Security Council. He visited both the countries and came up with a plan of demilitarization for the state of J and K, in addition to the problems of northern areas of the state and the disposition of the Azad forces (United Nations Official Document Report 1948).

Mcnaughton tries his best to solve the Kashmir dispute by suggesting demilitarization of forces from both sides. But India rejected its proposals by contending that Kashmir had accessed to India and as such it becomes a part of Indian territory and the defence of which is the responsibility of the Indian forces (Das 1950). Hence, the mediation of Mcnaughton failed and the issue again came before the Security Council.

Mediation Efforts of Sir Owen Dixon

The United Nations Security Council adopted another resolution on March 14, 1950 by appointing Sir Owen Dixon as the UN representative for India and Pakistan. He was also entrusted with the plan of demilitarization for the state of Jammu and Kashmir. After visiting both the countries, he recognized that India did not claim the right of Azad Kashmir forces in the state and also the military forces allotted to her in his proposal for demilitarization were unrealistic (Brown 1996). The Pakistan on the other hand argued that it is a breach of India's agreement and though refused to accept his proposal. Finally, Dixon submitted the report on 15 September, 1950 and recommended that partition of the state would offer the best possibility for settling the problem (Ibid).

Mediation Efforts of Dr. Graham

On April 30, 1951, the UN Security Council appointed Dr. Graham as the UN representative for the Kashmir dispute. He also worked on the same demilitarization agenda of the state. Though, he submitted four reports to the United Nations Security Council which were every time rejected by India. But in his final report, he informed the Security Council on 23, January 1953 that both the parties have agreed to resolve the conflict by agreeing on the reduction of troops from both sides (Brecher 1953). He suggested the number and character of forces which should remain at the end of demilitarization in the state of Jammu and Kashmir. On the Pakistani side, an armed force of 6,000 separated from the administrative and operational command of the Pakistan without armour or artillery; while as on the Indian side, an armed force of 21,000 including state armed force with no armour or artillery. This time India accepted the figure proposed, but at the same time rejecting the presence of 6,000 armed troops in Azad Kashmir, along with removal of all Pakistani appointed officials of Azad Kashmir. This approach of India makes Pakistan more critical and with this disagreement, the mediation process of Dr. Graham ended. However, he recommended the leader of both the countries to undertake the direct negotiations to resolve the impasse (Ibid).

From the period 1953 to 1957, the United Nations did not take part in the mediation process between India and Pakistan regarding the Kashmir dispute. However, the Security Council repeated its commitment of holding a plebiscite in the state of Jammu and Kashmir on 24 January, 1957. The president of the council-Gunnar Jarring was authorised to mediate between India and Pakistan on the proposals of demilitarization and plebiscite. Though, he failed to convince both the parties to accept the proposals but he referred a proposal of arbitration to the council. The proposal were again rejected by India and accepted by Pakistan. In 1958, Dr. Graham was again sent by the UN Security Council as a mediator between the two countries to secure their acceptance on the proposals of demilitarization and plebiscite, but he again failed in his mission. In March, 1958, he recommended to the Security Council to arbitrate the dispute, but India rejected the proposal of arbitration (Adhikari and Kamli 2010). Hence, the Kashmir issue finally died down at the Security Council because of India's unwillingness to follow the resolutions of UN.

To sum up, the United Nations have made several efforts between India and Pakistan to resolve the Kashmir dispute and passed various resolutions including Resolution on 21 April, 1948; Resolution on 13 August, 1948; Resolution on 5 January, 1949; Resolution on 14 March, 1950; Resolution on 30 March, 1951; Resolution on 24 January, 1957 etc. In all these resolutions, it was emphasised that Kashmir conflict should be resolved keeping in view the wish of the Kashmiris and the fate of Kashmir will be decided by the plebiscite. Besides, these resolutions also laid stress on resumption of peace and order, cease fires, negotiations etc. with the intention of eliminating the hostility, respect of LOC in order to create conducive atmosphere for talks and to create enviable and amity relationship. However, all the efforts made by the UN failed to provide any fruitful results and remained confined only to papers. With due course of time, Kashmir remained a contentious issue and the highly militarised place in the world. Besides, it always threatened security, peace and instability in the South Asian Region.

Kashmir: India-Pakistan Perspectives

The origin of the Kashmir conflict is rooted in British India's partition into the dominions of between India and Pakistan. It is assessed that the dispute generated out of the failure of the then British Government to implement the Government of India Act 1935 that predicted an establishment of federal system between the British Indian States and the princely states. Although, both the countries have intended many times to put an end to the conflict by diplomatic efforts, Confidence Building Measures, negotiations etc, but till now, they have failed to produce any fruitful gains. Moreover, the two countries have fought three wars over Kashmir, apart from several border skirmishes, but there is no end to the dispute.

On the one hand, the Indian Government contended that the Maharaja Hari Singh, the then ruler of the Jammu and Kashmir state signed an instrument of accession on 25th October, which was executed on 27th October with the government of India. It makes India a legal possessor of the state, the utility of which may be proved by India's presence in the state and its control over the territory. Also, the presence of military and armed forces in the valley prevailing there as a matter of right; the control of the defence, communication and external affairs of the state was a direct concern of the Act. Further, India maintains that it is reported in the United Nations Commission that "the cardinal feature of India's position is her contention that she is in Kashmir by right, and that Pakistan cannot aspire to equal footing with India in the contest" (Schofield 2000).

India considers the state of Jammu and Kashmir as an integral and inalienable part of the country and any question of negotiating on the issue of its accession is unrealistic. Besides, India views instrument of accession as final and cannot be disputed. If there is anything disputed or unfinished, it is the part of Jammu and Kashmir that Pakistan illegally occupied (Hussain 2009). In April 1948, the United Nations Commission on India and Pakistan adopted a resolution whereby, Pakistan had to withdraw its control from the Kashmir occupied part i.e, Azad Kashmir. Though, Pakistan never obeyed, India's stand is that the plebiscite (the second part of the resolution) could not have been followed. As such, it claims Kashmir as its own (Joshi 2003). India further maintains that Pakistan had lost the right to appeal the UN Resolutions, as it had made attempts to alter the status quo by force of war in 1965. In addition, India viewed that the greatest problem confronting the Kashmir is the terrorism that is sponsored by Pakistan and the sufferers of which are the Muslims in Kashmir (Hussain 2009).

Although, India claims Kashmir as its integral part, but at the same time use massive force to curb the Kashmiri's struggle of self-determination. In addition, India had always expressed a sort of seriousness in negotiating and discussing all outstanding issues with

Pakistan regarding Kashmir. But considering Kashmir as an integral part of it and thus moving away from its starting position. However, India is intending to maintain a strong position and hold over Kashmir and thus getting involved in dialogue process with Pakistan. At the international level, India is willing to make a conversion of Line of Control (LoC) into a permanent international boundary vis-a-vis locating the Kashmir struggle movement as a Pakistani sponsored terrorist activity.

On the other hand, Pakistan contended that the geographical proximity as well as the presence of majority of Muslim population in Kashmir makes it rightfully a part of Pakistan. It is based on the belief that Pakistan considers herself to be the only region in the subcontinent where Muslims could live outside the Hindu rule (Heymann 2003). Also, the partition of British India based on Jinnah's two nation theory was purely on religious lines and by following this theory, Kashmir should have been with Pakistan.

The Pakistan considers Kashmir as a disputed territory as the states accession to India in October 1947 was provisional and also its ruler was pressurised to accede to India. It considers the accession of the Jammu and Kashmir state to India as illegal as the Maharaja of the state had no authority to decide the fate of Kashmir. The people had already revolted against him and overthrown his government and had compelled him to flee from the valley (Schofield 2000). Taking somehow similar position, Alastair Lamb, the British historian adds Pakistan's statement on the instrument of accession, proved that the military intervention by India on 27 October 1947 took place before the Maharaja had formally acceded to India. Lamb maintained that there remains unambiguity whether the Maharaja ever did sign an Instrument of Accession as this document is said to be "missing" from Indian records. Lamb further shows that the record was falsified from the very beginning by claiming that a top Indian official, VP Menon, had travelled to Jammu on 26 October to secure such an Instrument, whereas Menon never visited Jammu on this date. Hence, the validity India's argument that it had sent its forces to defend its own territory had no basis since by October 27, as no Instrument of Accession had been signed (Lamb 1997).

Pakistan further viewed that the ruler of the state of Jammu and Kashmir had already signed a standstill agreement with Pakistan in order to maintain a status quo in the state, which India had violated by sending her troops in the valley. The Pakistan also maintained that the disputed status of the state of Jammu and Kashmir is recognised in the United Nations resolutions of August 13, 1948 and January 5, 1949 to which both India and Pakistan agreed and these remain fully in force today which cannot be disregarded by either party (Hussain 2009). Also, Pakistan contended that though, plebiscite never took place in the state of Jammu and Kashmir and hence, it should claim ownership of Kashmir (Joshi 2003).

India and the Kashmir Conflict

The Kashmir conflict can be viewed both from the perspectives of internal nuances within the Indian context as well as from the bilateral causes. In the internal context, it is related to the failures of Indian political system as it reveals the mounting and larger crises of India's political institutions while externally it is linked with the Indo-Pak rivalry. Also, the Kashmir problem is the result of lack of mutual trust and mutual suspicion on the part of both India and Pakistan. These countries have many occasions shown concern over the Kashmir conflict by means of negotiations but failed to provide any constructive gains. From the India's partition on 1947, the valley of Kashmir had witnessed political oppression as well as high handedness on the part of Indian government resulting in escalating the conflict (Bajpai and Ganguly 1994).

Kashmir Dispute in the United Nations: Indian Role

The Kashmir conflict remained a course of discussion in the United Nations for a long period of time. Primarily, both India and Pakistan accepted the option of plebiscite to settle down the Kashmir conflict. On the one hand, the government of India demanded an immediate withdrawal of Pakistani troops from its occupied territory. But, on the other hand, Pakistan while supporting the plebiscite, refused to withdraw its troops from PoK. Despite several efforts by the UN, both the countries continue to adhere to those positions and thereby discouraging multilateral resolution of the Kashmir issue (Ganguly 1990).

However, since 1948 the Indian government have challenged the validity of the resolutions of United Nations that have been once accepted by Indian government. Though, it was India that principally took the Kashmir issue to the United Nations Security Council on January 1, 1948. It launched the complaint against Pakistan under the violation of article 35 of the charter which may distort the international peace. It viewed Pakistan as an aggressor and wanted UN interference for the peaceful settlement of the dispute. With this complaint, the UN mission arrived in Kashmir and passed many resolutions. But India imparted reluctance every time and its attitude was negative instead being the initiator, did not follow the recommendations of the UN (Hilali 2007). Besides, the Indian Representative -Gopalaswami Ayyenger told the UN Security Council on January 14, 1948 that "the question of the future status of Kashmir vis-à-vis her neighbors and the world at large, and a further question, namely whether she should withdraw from her accession to India, and either accede to Pakistan or remain independent, with a right to claim admission as a member of the United Nations - all this we have recognized to be a matter for unfettered decision by the people of Kashmir, after normal life is restored to them" (Amin 2003).

Moreover, the Indian government never acknowledged any suggestion regarding demilitarisation and conducting a plebiscite in Kashmir. In this perspective, India's attitude towards the conflict is politically indefensible and morally bankrupt. Also, it never accepted any proposal concerning demilitarisation and holding a plebiscite in Kashmir. In addition, India had no hesitation in breaking the promises and commitments vis-à-vis claiming that the resolutions passed by the UN had lost their relevance. Instead, the UN resolutions of August 13, 1948 and January 5, 1949 determine the nature of the Kashmir conflict. Due to lack of political will to resolve the Kashmir dispute in the United Nations, the valley of Kashmir has been a source of instability, depression, tension and bloodshed in the south Asian region (Hilali 2007).

As India's disillusionment with the UN sponsored negotiations grew, it gradually moved to extend its administrative control over Kashmir. In this context, the local elections were held in 1951, providing Sheikh Abdullah's Kashmir National Conference an overwhelming victory (Ganguly 1990). But it is to be pointed out that Indian government denied the people of Kashmir from their basic rights and freedoms time and again, which leads towards escalating the conflict. Thus, the matter of fact is that the Kashmir issue is still one of the contentious and biggest unresolved issues in the United Nations.

India and the Political Process in Kashmir

After the accession of the state, Nehru the then prime minister of India wished the Maharaja Hari Singh to quit and transfer the power to sheikh Abdullah – the popular Kashmiri leader. The reason behind being was two folded. The first was the clause 7 and 8 of the accession[3] and the second was the Abdullah's hatred against the maharaja. Nehru contended that the complete involvement of Sheikh Abdullah was essential for both national and international reasons. It was within this context that the Abdullah occupied its first political role as head of emergency administrator for the state of J and K. Finally, on March 2, 1948, Sheikh Abdullah was appointed as the prime minister of the state.

Although, the Jammu and Kashmir constituent assembly witnessed its first ever elections in September 1951. The national conference as being single contester won all the 75 seats (Bhatti and Bakaya 2005). However, Sheikh Abdullah remained firm that the Indian government had no right to extend its jurisdiction in Kashmir beyond the three realms as mentioned in the instrument of accession i.e, defence, foreign affairs and communication. It was under such circumstances that article 370 of the Indian constitution was granted to the state of Jammu

---

[3] Clause 7 of the accession states that nothing in this instrument shall be deemed to commit me in any way to acceptance of any future constitution of India. While as Clause 8 of the accession reads that nothing in this instrument affects the continuance of my sovereignty in and over this state.

and Kashmir[4]. In July 1952, he signed Delhi agreement[5] on a number of issues with the Indian government. However, he was dismissed as a prime minister on august 8, 1953 and was put behind bars by the Indian authorities. The Indian government charged him on the grounds that he was acting against the interests of the Indian Union, was corrupt and was looking for independent status of the state. On his dismissal, Abdullah states in his book that "how did a patriot, praised by mahatma Gandhi and Jawaharlal Nehru for his straightforwardness, turn into an enemy of the country?" It may be viewed here that Abdullah's delay in approving the instrument of accession was not fulfilling India's objective of strengthening its position on Kashmir (Schofield 2000).

The dismissal of Sheikh Abdullah resulted in widespread protests throughout the valley leading to the death of sixty Kashmiris (Malik 2002). His position was occupied by his Deputy - Bakshi Gulam Muhammad. The reign of Bakshi which lasted for ten years were the years of dictatorship and corruption. During his tenure, the people in the state of J and K were denied from basic civil liberties. Even numerous newspapers were banned who had criticised the government including the 'voice of Kashmir'. It can be said that people enjoyed more freedom under the Maharaja than that of Bakshi's government. According to the ICG Asian

---

[4] Article 370 of the Indian constitution grants special status to the state of J and K. This article authorises J and K to have its own constituent assembly and a separate constitution. Also, the Indian government has no power to apply proclamation of financial emergency in the state of J and K. In addition, the parliament cannot make laws on the state list and also, the jurisdiction of union parliament cannot be extended to the state of Jammu and Kashmir. Further, article 368(i.e, amending the constitution) cannot be applied to the state and also the residuary powers rest with the J and K state.

[5] Delhi agreement states that the residuary powers rest in the state itself. Also, the state would have its own flag. In addition, the supreme court of India would only have appellate jurisdiction on the state of Jammu and Kashmir.

Report (2003), this period witnessed a widespread dissatisfaction with the state's new puppet government resulting into the random violence for the ten years including riots in 1955 and 1957.

During Abdullah's trail, elections were deliberately arranged in a way as to prevent pro-Abdullah from winning. It is evident that Nehru in 1962 pointed to Bakshi Gulam Mohammad that "your position would be strengthened much more even if you lost a few seats to bona fide opponents" (Bose, et.al 2000). However, it is observed that India fulfilled most of its objectives regarding Kashmir during those ten years as:

- The constituent assembly of the state of Jammu and Kashmir was ratified in February 1954[6].
- In January 1957, the Jammu and Kashmir's constituent assembly adopted the new constitution[7].
- In February 1958, the powers of Comptroller and Auditorial general of India were extended to the state.

However, in 1963, Bakshi Gulam Mohammad was dismissed and was arrested under the Defence of India Act. He was replaced by Shamsudddin, but after few months, he too was dismissed and G. M. Sadiq was brought in (Bose, et.al 2000). In his regime too, several other provisions of the Indian constitution were extended to the state of Jammu and Kashmir.

---

[6] In 1954, by the presidential orders under article 370, the preamble and other provisions of the Indian constitution such as provisions relating to union and its territory citizenship, fundamental rights, part xi and part xii relating to finance and part xiii were extended to the state of J and K. in addition the customs and duties to be levied by the state of Jammu and Kashmir itself were abolished.

[7] By adopting the new constitution, most of the provisions of the Indian constitution were extended to the state. In addition, the Indian government declared that the instrument of accession had been decided once for all. However, it may be mentioned that during this period Sheikh Abdullah and his colleagues remained in jail.

Article 249 of the Indian constitution was the first of this kind, under which the centre was authorised to enact laws on any matter mentioned in the state list. Besides, Articles 356 and 357 under presidential orders were also extended to the state on December 21, 1965[8]. Moreover, the 64th Constitutional Amendment was introduced which relates to the change of Sadri-Riyasat to the Governor and the Prime Minister to the Chief Minister, along with bringing the judges of the High Court of J and K at par with the judges of High Court in all other states of India.

Meanwhile, Sheikh Abdullah was released from jail in 1968 and on his return to Srinagar, he was widely praised and admired. With the course of time, his party's membership grew. To keep Abdullah under check for his political vision of Kashmir, Indira Gandhi, the then prime minister of India pressurised him to form an electoral agreement with her state Congress party in the 1972 elections. He refused her offer which led him to renewed detention in 1971 and his party (Plebiscite Front) was banned to participate in the elections, along with declaring it as an illegal organisation (Schofield 2000). In 1972, the congress party led by Syed Mir Qasim won the elections and formed the government in the state.

Finally, in June 1972, Sheikh Abdullah dropped his plebiscite demand and changed his stance by contending that our dispute with the Government of India is not on the question of accession, but on the proportion or quantity of autonomy. This led to a series of negotiations, which paved a way towards signing of 'Kashmir Accord' on 12 February, 1975 with the government of India (Adhikari and Kamle 2010). By this

---

[8] Article 356 empowers the president to assume the functions of the government of the state in case of failure of constitutional machinery, while as Article 357 empowers parliament to exercise legislative powers of the state of Jammu and Kashmir.

accord, Sheikh Abdullah was to accept that the state of Jammu and Kashmir was an integral part of India. In return, he was assured that all the Ordinances and Acts were to be reviewed which were issued after his arrest in 1953 (Navlakha 1991). After five months from the accord, the election to the State Assembly was to be held in July 1975. The national conference won the elections and he became the Chief Minister of the state. He governed the State until he died on September 21, 1982.

In 1983, the elections for state assembly were held. The elections were won by National Conference led by Dr. Farooq Abdullah (son of Sheikh Abdullah). After forming the government in the state, Farooq demanded for the restoration of the pre-1952 status for J and K state leading to his dismissal in 1984 by the government of India. After his dismissal, the State of Jammu and Kashmir was brought under the Central Rule for a brief period and a puppet government under G. M. Shah was formed by the Government of India (Adhikari and Kamle 2010). However, following the Hindu-Muslim tension in the valley, this government was also dismissed by the central government in 1986 and the State was again put under the Central Rule. Though, surprisingly in June 1986, Farooq Abdullah signed a deal with Rajiv Gandhi to form a coalition government with the congress in the state. In March 1987, the elections were held in the state and the coalition of the Congress-National Conference alliance won. But the victory of the alliance was perceived to have engaged in electoral fraud. The hijacking of the political process in the state leads to discontentment among the people in the valley. This political failure aroused people in the valley and they were largely drawn to insurgent and separatist movements (Bajpai and Ganguly 1994). Hence the valley witnessed the rise of insurgency in early 1990s (since 1989) mainly due to the malpractices on the part of government of India.

Since the integration of the state of Jammu and Kashmir with the Indian Union, there was discontent particularly in the valley mostly after the ratification of the instrument of accession in 1954. The mistrust concealed by the government of India towards National Conference as well as forcing state's merger with India, resulted to the erosion of internal autonomy of the state of Jammu and Kashmir. Also, most of the provisions that took place in 1960's like the extension of the jurisdiction of Comptroller and Auditorial General of India, election commission, Supreme Court and a series of central laws led to further disillusionment. The state of Jammu and Kashmir witnessed the denial of democracy where the centre frequently dismissed Kashmiri leader in order to administer its "will" in the state. In addition, the 'Kashmir Accord' signed by the Sheikh Abdullah with the government of India was never put before parliament. It was in this context that the Sheikh Abdullah requested that his body should be taken to the Arabian Sea, rather than to be buried in Kashmir, as he considered 'Kashmir as an enslaved territory' (Bose, et.al 2000).

Generally, the setting up of puppet and repressive regimes through undemocratic methods, imposing president's rule, erosion of autonomy under article 370 of the Indian constitution, interfering in state matters, opportunistic accords and alliances with the National Conference, electoral malpractices and using of armed forces to suppress agitation had greatly led towards escalating the Kashmir conflict.

Militarization in the Kashmir Valley

The valley of Kashmir is one of the world's most heavily militarized places where special laws are currently promulgated. The parameters of militarisation in the valley of Kashmir are established by a series of emergency provisions like the Armed Forces Special Powers Act (AFSPA), Public Safety Act (PSA), Terrorist and Disruptive Activities Act (TADA), Disturbed Areas Act (DAA) and the Prevention of Terrorism Act (POTA). As a result of these special powers and the impunity granted under these acts, thousands of people were killed and migrated, the livelihoods were destroyed and the people were forced to live in an atmosphere of uncertainty and fear (Qutab 2012). Under the Terrorist and Disruptive Activities Act of 1987, a person is denied from basic rights such as the right of the accused to be informed of the charges against him. The act also permits even low-ranking officers in the civilian security forces the right to use violence without the permission of higher authorities (Jones 2008).

Apart from TADA, the most controversial law prevailing is the AFSPA which grants the armed forces some extraordinary powers like the power to arrest and enter property without warrant and to fire upon or otherwise use force, even to the causing of death, against any person who acts contrary of any law or order if the armed personnel believes it necessary for the maintenance of public order. The intensive militarisation in the valley of Kashmir authorised by the AFSPA establishes conditions of a legal civil war against civilians and in practice dominates the administration of everyday life in the state. Also, the military ethos and values through this act have become dominant throughout various structures and institutions of the state like the judicial process, law enforcement and civil society. Under the AFSPA, the Kashmiris live in a state of constant fear of arbitrary arrests, sexual harassment, enforced disappearances, torture, and custodial deaths vis-à-vis have become the permanent prisoners of the state. By the prevalence of these cruel atrocities, the Kashmiris refer to their homeland as an occupied zone, a garrisoned state, a bunkered territory, and not an occupied territory (Duschinski 2009). In addition, the PSA which is only applicable to the state of Jammu and Kashmir and under which a person can be detained for the period of two years without any trail vis-a-vis the detaining authority does not disclose the facts of detention. Under this act, estimates of the number of detainees in the last two decades could range from 8000- 20000 (Singh 2011).

With the intensive militarisation as well as in the presence of these acts, it is estimated that over the past two decades, a death toll of over 70,000, detention of more than 60,000 civilians; torture of nearly 20,000 detainees and enforced disappearance of 8,000 have been witnessed in the Kashmir valley (Navlakha 2008). Also, there are around 63 interrogation centres in Kashmir run by the military and para-military forces where torture has been carried out. The army camps, hotels and several other buildings have been taken over by the armed forces as detention centres (Schofield 2000) vis-à-vis continues to occupy civil buildings like the migrant houses, hotels, office buildings, industrial areas, cinemas, college hostels and so on. These interrogation centres remain beyond the judicial scrutiny and thereby making it nearly impossible for the Kashmiri people to ascertain the whereabouts or welfare of those who have been in custody or have simply disappeared (Kazi 2007). Under AFSPA, Kashmiris have become permanent prisoners of the state and through intensive militarization of Kashmiri society, India administered Kashmir valley through a form civil military authoritarianism (Duschinshki 2009).

As effect of militarization, massive human rights violations are taking place in the valley of Kashmir. Due to this violent conflict, thousands of civilians have died in regular clashes between separatist militants and security forces. The Srinagar based international people's tribunal on human rights and justice in Indian administered Kashmir and the association of parents of disappeared persons prepared a 354 page report mentioned in international business times (7th Dec, 2012), on human rights violations. The report said that an estimate of 8000 persons had disappeared in the past two decades, in addition to the death of more than 70,000 people. It cited disclosers of more than 6000 unknown, unmarked and mass graves.

The militarization in the Kashmir valley has grossly undermined the ethos of democracy in which human rights violations seems to be day today affair. Worse many civilians have to bear the cost of losing near and dear that would make their lives harder to live. Thus, it is hard to conceive life of human being under constant fear and having in position to realise their potentiality. What could one say but reminded by a famous poem of Faiz, whom ever you want to blame, blame it.

Protests in Kashmir

The valley of Kashmir remained tense and disturbed due to the occurrence of violence since early 1990s. However, from 2002, the violent incidents in the Kashmir valley had considerably declined due to reduction in militancy wave but, the valley witnessed civil unrest in 2008 after the government decided to transfer forest land of 100 acres to Shri Amarnath Shrine board in Phalgam area of District Anantnag. The state was then ruled by a coalition of Congress and People's Democratic Party (PDP). But the PDP by opposing the transfer of land withdraw its support from the coalition after the crises erupted in the valley (Singh 2011). Soon after the news of land transfer spread across the valley, the people raised protests against the government. Apart from Srinagar, the protests were reported from all parts of the valley including Anantnag, Kulgam, Pulwama, Budgam, Baramullah, Bandipora, Ganderbal, Shopian and Kupwara.

According to BBC News (2008), the people in the valley launched heavy protests against the allocation of land by viewing that it was intended at shifting the demographic equilibrium in the area. On the other hand, the state government claimed that the land was needed to build sheds and toilets for the pilgrims visiting the Amarnath shrine. The violent conflicts between the people of valley and police-paramilitary forces continued for 61 days until the government revoked the orders. However, the agitation resulted in the loss of 63 people as reported by the J and K State Human Rights Commission (Free Press Kashmir 2008).

The Kashmir again witnessed the civil unrest in June 2010, after the killing of two civilians by the security forces (Al Jazeera 2010). The people in the valley protested against the human rights violations prevailing in the valley. To normalise the situation, the government imposed curfew and deployed thousands of police and paramilitary forces. However, the security forces beat people while attending the funeral possessions which increased anger among the people resulted in escalating the agitation. Thousands of protesters from all areas of the valley joined the protests. These protests between the civilians and the police-paramilitary forces lasted more than three months resulting in the death of over 200 civilians (Singh 2011).

On 19 July, 2013, the valley again erupted in agitation against the desecration of the holy Quran in the district of Ramban (J and K). The local inhabitants protested against it which caused the death of 4 civilians by the BSF personnel who opened fire on them (Economic Times 2013). Another agitation was witnessed in the valley on 8 September, 2013, against the Zubin Mehta's music concert.[9] The protests broke out in the valley resulting in the death of five civilians by the security forces (CRPF) in the Shopian district. (Times of India 2013). In response to these killings and incidents, the separatist leader-Syed Ali Shah Geelani said that "It is imperative to fight oppressive forces and we have no option other than to strive for freedom from New Delhi's forcible control". He further added that such incidents would continue to take place unless the forced occupation remains in the state (Gul 2013).

The process of innocent deaths of civilians continued every year till the year 2016 in which the protests assumed a mass dimension. In 2016, the Kashmir witnessed civil unrest over the killing of Burhan Wani, a Hizb militant commander by the forces on 08 July. After his killing, anti-Indian protests aroused across all 10 districts of the Kashmir. As a result, mobile services were suspended and curfew was imposed in entire valley for 53 consecutive days, however curfew was reimposed in some areas the next day. The paramilitary forces and Jammu and Kashmir Police used pellet guns, pava shells, rubber bullets and tear gas shells resulting in death of more than 90 civilians and over 15,000 injured (out of which over 750 injured civilians were hit by pellets in eyes, with many of them facing blindness, another 343 persons suffered bullet injuries and more than 4500 persons were treated for pellet injuries in body parts apart from eyes).

---

[9] Kashmiri's viewed the music concert as a cultural violation.

Besides, more 12,000 were arrested and detained and hundreds others were slapped with the PSA (Rather 2017).

Thus, while analysing the Kashmir conflict, it becomes evident that the protests largely reflect a shift in the repertoire of Kashmiri resistance from armed militancy to contentious politics, with thousands of people taking to streets for a prolonged period. This shift in the repertoire of resistance is significant, giving the people of Kashmiri a tool of disruption and attention that cannot be dismissed as simply "Pakistan backed terrorism".

Role of Pakistan in Kashmir Conflict

The Kashmir conflict was born in 1947 at the time of partition of British-India into the dominions of India and Pakistan. The conflict is both a cause as well as consequence of the Indo-Pak relations. It is principally a conflict about justice and human life, while its territorial and strategic dimensions are adequately difficult. Both the countries wished to join the state of Jammu and Kashmir to their own sides for their respective interests. The Kashmir conflict has been less competing over strategic or resource grounds but more over visions of state building and nationalism (Ganguly and Bajpai 1994). The nation building strategy of Pakistan viewed Kashmir as an integral component to be added in its own ambit. During the time of independence, the maharaja of Kashmir had signed a standstill agreement with Pakistan that allows supplying essential commodities like gasoline, wheat, cloth etc in the state. Later on, the government of Pakistan violated the agreement by stopping in delivering the commodities (Thornier 1949).

The former was followed by the invasion on the part of well-armed Pathan tribals from the Northwest Frontier Province (NWFP) of

Pakistan that is often noted as first sign of conflict in the valley of Kashmir. In the meantime, the ruler of Kashmir requested Indian government to help so that invaders would be driven back. Indian agreed upon help with the condition of signing the instrument of accession. After agreement, the Indian army released the valley from the grip of invaders and perpetuators of violence (Ashutosh 1991). However, Pakistan runs on the risk of putting the strategy of liberation in the valley of Kashmir that would inflict tension in other parts of India, especially more than 100 million Muslims.

Coming back to invasion, Pakistan denied its involvement, even though it is clear-cut of having internal support that helped in carrying out violence. On 1st January, 1948, India refers the issue of Kashmir to the United Nation which noted that there is visible presence of Pakistani military. However, it recommended twofold strategies to mitigate the tension: 1) the withdrawal of Pakistan tribesman as well as national of Pakistan from the territory of Kashmir including Pakistan Occupied Territory (POK) commonly known as Azad Kashmir, and 2) Simultaneously, the withdrawal of bulk military from Kashmir. If both countries would agree on the above recommendations, only then a plebiscite might have taken place in Kashmir. But, Pakistan could not trust India that led to the presence of military even after UN recommendation. Also, Pakistan's interest solely lies in incorporation of Kashmir into its territory but not in independence, even if the people of Kashmir wish (Ibid). The reluctance imparted by the Pakistan regarding UN recommendations has contributed greatly in escalating the tension and the direct outcome is the sufferings of people of valley even witnessing today. In addition, the tendency of Pakistan to canonize the voice of people from Kashmir has been resulted in prolonging the tension.

In the early 1970s, Pakistan initiated assisting and providing support to the militant actions in the state of Jammu and Kashmir with the aim of liberating Kashmir. The first militant activity witnessed in the valley was hijacking of Indian aircraft in 1971 and 1976 respectively which was operated by the Jammu and Kashmir Liberation Front (JKLF). The organisation received proper funding, weapons and training facilities from the state of Pakistan. Besides providing support, the intelligence agencies of Pakistan such as the Field Intelligence Unit (FRU) and Inter-Services Intelligence Directorate (ISI) started getting involved in providing arms training to Kashmiri youths (Singh 1995). The aim behind sponsoring militancy in Kashmir was a part of Pakistan's strategic long-term programme to reveal its loyalty towards Kashmiri's and thereby gaining ground to maintain its interest in the state mainly in the valley. However, the role of Pakistan was limited in starting the uprising in the valley but has been a critical factor in sustaining it. (Cohen 2002).

In early 1990, there were as many as 30 militant groups active in the valley of Kashmir sharing various ideologies (Singh 1995). Most of these groups received shelter and assistance from Pakistan. The most prominent among them were Hizbul Mujahideen and JKLF. The former wished Kashmir to be part of Pakistan while the latter wants an independent Kashmir (Varshney 1991). These groups received political support from the state of Pakistan and the activeness of them in the valley created a troublesome environment leading to vast bloodshed of the Kashmiri. According to the report of Asia Watch, militant groups prevailing in the valley have committed grave human rights violations. In several cases, these groups have deliberately targeted civilians causing death of hundreds of civilians and the persons associated with political groups. In addition, nearly 100,000 Hindus as well as 10,000 Muslims have been migrated from the valley due to threats posed by these groups. There is no clear-cut figure about the number of civilians killed by the militants but the government sources reveal that more than 2,000 civilians have been killed by the militant groups.

Hence, the involvement of Pakistan through various means to bring Kashmir under its ambit is undisputable and its inspiration has been resulted in loss of innocent lives as well as gross violation of human rights. Besides, the voices of Kashmir on the demand of being independent have been undermined. In addition, the political instability in Pakistan explicitly (or implicitly) has brought the tension in the fore which further derailed the chances of peace negotiation between the two countries.

To sum up, the genesis of Kashmir conflict could be viewed by considering various factors like the failure of British government to provide a peaceful, logical and politically accepted partition, political rivalries between India and Pakistan, importance of Kashmir in terms of national identities of both countries and the short-sightedness of political leaders. Also, the ineffectiveness of foreign policy of both the countries to resolve the tension might have mitigated the hostility between them.

CHAPTER – 4
CONTEXTUALISING HUMAN SECURITY CHALLENGES IN KASHMIR VALLEY

The conception of human security combined positive and negative liberty in broader sense. Thereby, placing individual at the centre of discussion, that is quite contrary to the traditional or conventional theories of security in which nation states are focal point of debate. As previous chapter pointed out that human security encompasses freedom from fear and freedom from want, namely fulfilment of basic needs in order to evaluate how and what extent human being's freedoms are realized. Nevertheless, the conception of freedom is such that bound to create such deeper division among scholars over what constitutes freedom and might goes to hypothetical state of affairs. But there are fruitful grounds to have an agreement on the path of realizing individual's life and which way it could be enhanced. For instance, there might not be realization of freedom if one is deprived from food or inability to participate in decision making of which affect their lives directly or indirectly.

The assessment and evaluation in accordance with human security has been conducted in different parts of the globe in which the continuity or discontinuity between and within local, national and international security is explored. For instance, UNESCO studies for Arab states (2005), Caribbean and Latin American (2005), East Asia (2004) and Central Asia (2006) (Acharya, et.al 2011). Against this backdrop, Acharya, et.al, developed a human security governance index that measures challenges to human security and responses to human insecurity. That index would do some ranking whereby prioritisation could be driven (Ibid: 2-3).This is sustentative with mapping out human security in the North Eastern Region of India including Orissa. Nevertheless, the extension of human security conception in the valley of Kashmir has not attracted the attention of scholars. Since, it is well known that the valley has been the conflict prone region, given its historical roots lie in India-Pakistan relationship and intensified by militancy-militarization dichotomy.

Since the inception of India from British colony as an independent state, one of the major region namely Kashmir valley has been conflict prone area within national boundary as well as at the international level. Conventionally, Kashmir has been viewed from bringing peace through military intervention. But more than six decades of independent India suggests that development or underdevelopment of Kashmir persist or even intensify. Balraj Puri argues that looking back objectively, one can trace the beginning of Kashmir problem and its aggravation to the denial to the people of state of civil liberties, of democratic and human rights including the right to freedom of speech, of the right to protest and form an opposite party, of the right to vote and to elect a government of their choice (Puri 1999). It is not enough to locate the Kashmir problem in one dimensional way but demands multiple directions. For instance, like of economics opportunity, persistent poverty, and high rate of illiteracy, environmental degradation and so on has some way or the other way kept Kashmir in the trap of underdevelopment. Considering the above-mentioned factors, there is a need to contextualise the very development or underdevelopment of Kashmir from socio-economic angle to check the life of people in the valley.

Socio-Economic Factors

The Kashmir valley has observed severe challenges in all dimensions be it the economic, political and socio-cultural, due to the persistent violence and distressing environment. Under such an atmosphere, the people remained in the state of distrust which destroys their civic life and leads to stunted economic growth. The Kashmir problem is identified as a complex situation witnessing the conflict between the two nuclear powers and on the other side, the people developing alienation against the Indian federal rule (Schofield 2000). This conflict in the valley created a scenario which led to the stagnation of economy not only to the particular sector of investment or industry. It had also discouraged private investment apart from affecting the basic sources of livelihood including handicrafts, tourism and horticulture. In addition, it had caused heavy damage to the infrastructure and the outcome of all this being the common people who suffered the most (Evans 2000).

Poverty

Poverty has been conventionally defined as lack of resources at certain threshold leading to the deprivation. It could be seen in the 'Basic needs conception' where the direction has been taken to identify that what could constitute basics requirement of human in order to be out of deprivation, but critics point out that even person with having the resource disposal at hand might not able to realize it in the direction of own wishes. Thus, it needs extra resources in order to combat the deprivation. One may think of a household with the adequate food availability but girl child is undernourished. In such case one has to step beyond the basic necessity to conceptualize the deprivation and poverty is multi-layered and multifaceted deprivation. Any ways, the poverty is defined on two broader grounds, namely relative as well as absolute, and many reports take the later conception of poverty in which identification is based on certain level of threshold to be as minimal to meet in order to have out of deprivation.

Poverty is viewed as a complex socio-economic status resulting out of economic inefficiency and a condition of political and social exclusion. It has been observed that poverty leads towards malnutrition and hunger, disease and even to conflict or civil war. The intensive militarisation and the armed conflict in the valley had also increased the poverty leading towards malnutrition and hunger, thereby, creating issues of food security. According to the economic survey (2012-2013) made by the J and K government, the state has a population of over 24 lakh persons falling below the poverty line category. In comparison, the total estimated Below Poverty Line (BPL) population ratio in Kashmir valley has been nearly 13 lakh persons (12.91) i.e, 21.37 Percent. Opposing to the assertions of state government of making all round development in the state, every fifth citizen of the valley is falling under the BPL.

The report further showed that the Kashmir valley had more BPL population ratio at rural level as compared to the Jammu division. It also revealed that the state lagged behind the national average in poverty reduction in the span of 10 years. From years 1993-94 to 2007-2008, the national level of poverty has declined by 23 percent as compared to J and

K which has only shown a decline of 14.06 percent. The report also indicated that the three districts of the valley has highest BPL population including Baramullah with 2.37 lakh poor person, Kupwara with 2.34 lakhs BPL population and Budgam with estimated BPL population of 1.78 lakh persons as compared to Leh with only 0.29 lakh persons under BPL population.

Unemployment

The Human Security conception does see the unemployment as the greater causes of deprivation in which people might not able to live their life as they wish to, as having ability to the purchasing power along with a sense of fulfilment to be engaged in productive activities. Broadly, the unemployment is defined a person is seeking job but unable to get in certain time lag e.g, week, or month. Since long, the state of J and K particularly the valley has been facing the problem of unemployment. Due to the absence of desirable industrial growth and the inadequate opportunity of absorption in the private sector, the valley had witnessed alarming unemployment problems.

According to the economic survey 2012-2013, the number of unemployed youth of the state of J and K was 5.94 lakhs by ending of September, 2012. However, there are 3, 21,526 unemployed youth in the valley of Kashmir as compared to 2, 80,285 in the Jammu province. Comparatively, the state has highest unemployment of 5.3 percent among northern states. At the all India level, the figures for the unemployment rate are only 2.6 percent. (Kathju 2013). But the ranking of literacy in the state had improved from 32nd position in 2001 to the 30th position in 2013. On the other hand, the rate of employment had shown a reverse trend is being increasing alarmingly (Raj 2013). The valley of Kashmir in the absence of solid economy lacked the support in terms of job markets. As a result, the qualified and educated individuals confronted with scarcity of employment with respect of their relevant jobs. It is estimated that the average per capita income in the valley of Kashmir is 100 times less than the United States at about $ 325 a year (Arshad 2008). Also, the problem of unemployment could be attributed as a cause to the crises in Kashmir valley as expressed by the former Union Minister and senior Congress leader Karan Singh (Financial express 2013).

This unemployment further resulted in un-meeting of normal social life pushing towards further poverty. Also, thousands of low level educated students with little or no hope for a job drive themselves into terror groups while as thousands of local boys due to the hardships created by military forces turned to militant training across the border (Islam 2014).

Agriculture and Environment

The economy of India is still defined by the largest presence of people in the agriculture, so is the case with Kashmir valley. The agricultural sector needs to be given priority for ensuring inclusive growth for reducing poverty growth vis-à-vis providing food security to the people. Since, the largest number of population in the valley remained engaged in agricultural activity; the prospects seem to be dismally lower as compared to other state of India. In the state of J and K, nearly 70 percent of the population gets their livelihood directly or indirectly from the agriculture. However, the contribution of agriculture towards gross state domestic product (GSDP) has gradually decreased as it was 21.19 percent in 2011-12 as against 28.06 percent in 2004-05 (Economic Survey, J and K 2012-13).

The agriculture in the Kashmir valley as being the major source of income suffered several setbacks and remained poor due to low productivity, lack of adequate irrigation facilities (58 percent of net area sown is rained) and want of adequate agricultural research. Apart from it, there are inadequate and unorganised marketing facilities and distant markets for export outside the valley. Besides, the heavy militarisation in the valley since the beginning of armed conflict has also usurped agricultural lands and residential areas as well as has taken a heavy toll of the forests. The presence of military and paramilitary forces have established their camps in forests in order to target militants hiding in the area in which the wildlife is becoming target of the combat. Thus, the human encroachments and operations launched by the military forces led to the destruction of these forests by which state gets deprived of many plants having the medicinal value.

Moreover, the armed conflict has greatly troubled the flora and fauna of the region due to the huge pressure of the armed forces on the glaciers. Moreover, the Indian forces have often occupied and established their camps nearby the water bodies in the valley. The combined effects of deforestation and destruction of water resources has led to soil erosion which is responsible for frequent flash floods now visible in the state of Jammu and Kashmir (Rehman 2009). Not only the deforestation but the valley has also lost the vast wildlife reserves that provided habitat for some of the rare and endangered species like the snow leopard, antelope, Hangul, blue sheep, black bear and musk deer. In addition to this, it is estimated that over hundred varieties of birds such as quills, eagles, kites, pheasants etc. have been virtually disappeared due to the loss of natural habitat. Similarly, the wetlands, springs, lakes (like Dal Lake, Wular Lake) and water bodies (like Jehlum) have become the victims of apathy (Mahapatra and Shekhawat 2008).

Industry

Industry is the backbone of development as well as prosperity for any state. However, J and K particularly the valley lagged behind in this sector to attract investments and thus remained an industrialised backward region. The valley remained economically disadvantaged because of its poor connectivity and remoteness, hilly terrain, poor infrastructure and weak resource base. Also, the valley is having a disadvantageous position in terms of industrial growth and progress and the major industries namely the tourism and handicraft suffered heavy setbacks by the armed conflict in general and intensive militarisation in particular. Moreover, the political uncertainty and prolonged economic shutdown had adversely affected the pace of industrialisation in the region (Economic Survey, J and K 2012-13).

Besides, the tourist spots located in the valley were either partially or totally encroached by army pickets and camps. Also, many of the important historical sites were occupied by Indian military and paramilitary forces including the world famous Mughal gardens. The tourist spots under the occupation of the Indian forces have kept away the local tourists from visiting such destinations. In addition, the tourists were also targeted by various militancy related incidents. Prior to violence, tourism accounted for nearly ten percent of the state's income but during the years of unrest, tourism contributed virtually nothing to the state's economy (Shah, Wani and Mustafa 2013). Thus, due to the violent conflict in Kashmir valley from over past 20 years, the region suffered severe economic setbacks. Even the important sources of livelihood of local people such as tourism and handicraft industries have been seriously affected. Though, in the absence of heavy industries, the valley witnessed impediments also in the large and medium scale sectors.

(a) Handicrafts

The valley of Kashmir enjoys a rich history of handicrafts and is known as the home of handicraft goods. For decades, this sector has been the backbone of economy of the valley. The handicrafts have been substantial in fulfilling the daily requirements of local people and in serving their livelihood. Also, they are very renowned and have great economic potential in exports. But instead of enjoying worldwide name and fame, their production has been suffered to a great extent due to the broke down of unrest in the valley since 1989. The table 4.1 on the next page shows the data relating to the Handicrafts year wise and their value of production:

Year	Actual value of production (Rs. in crores)	Year	Actual value of production (Rs. in crores)
1988-89	154.00	2002-03	775.00
1989-90	150.00	2003-04	821.53
1990-91	200.00	2004-05	887.00
1991-92	220.00	2005-06	900.00
1992-93	230.00	2006-07	950.00
1993-94	240.00	2007-08	1614.59
1994-95	245.00	2008-09	1100.00
1995-96	250.00	2009-10	1000.00
1996-97	260.00	2010-11	325.00
1997-98	260.00		
1998-99	280.00		
1999-00	633.03		
2000-01	696.33		
2001-02	765.94		

Table: 4.1. Year wise production of Handicrafts and their value

Source: Directorate of Handicrafts (J and K Government).

Thus, handicrafts in valley faced several limitations with the turmoil whereby its market remained confined and could not integrate with the national and global economy (Ahamad and Yasmin 2012). The handicraft production in the valley got deteriorated during 1989-99 and during the peak seasons of unrest in summer 2008-09, 2009-10 and 2010-11, the production got further deteriorated. The handicrafts besides being the important industry in the Kashmir valley, along with providing employment to more than three lakh people has lost its pace by the chaotic and violent environment in the valley. Besides, the day to day curfews, hartals and the confrontations between the armed forces and militants has greatly discouraged the growth of this industry. Recently, the World Bank has ranked the state of J and K at 30th position while evaluating the performance of economic reforms (Hassan 2016).

(b) Tourism

Tourism is one of the main industries in the valley of Kashmir and the great source of income for the local inhabitants. Also, the sector is boosted with great potentialities for the generation of employment even for the unskilled people. However, the sector had suffered greatly due to the violent environment and unrest prevailing in the valley. According to Mahapatra and Shekhawat (2008), there had been a substantial decrease in the number of tourists visiting the valley per year during the turmoil period from about 7, 000, 000 to only few thousand in the following years. The conflict had also cost a loss of $3.6 billion to tourism revenue. The table 4.2 on the next page shows the Year Wise Number of Tourist arrivals in the valley from 1987.

Year	No. of Tourists visiting Valley	Year	No. of Tourists visiting Valley
1987	721654	1999	217292
1988	722035	2000	111912
1988	557977	2001	72591
1990	10722	2002	27356
1991	6287	2003	191164
1992	10324	2004	376729
1993	8026	2005	603582
1994	9814	2006	432888
1995	8520	2007	441836
1996	9592	2008	22000
1997	16131	2009	20809
1998	109883	2010	24376

Table: 4.2. Year Wise Number of Tourist arrivals in the Valley

Source: Shah and et.al (2013)

It is estimated that the unrest in the state have resulted in the loss of 27 million tourists from 1989-2002. Due to the least tourist arrival in the valley, a large section of people associated with the sector changed their occupation during turmoil (Ibid). Thus, it became a great concern as a large section of people lost their employment and livelihood associated with this sector.

Health

Health is one among the components of human security and a fundamental human right. It is vital for individual's well-being as well as a prerequisite condition for economic development and growth at the national/state level. The promotion of good health improves the productivity and efficiency of the labour force and thereby contributing towards economic growth and leading to human welfare.

The valley of Kashmir is regarded by many as the paradise on earth. However, over the past two decades, it became associated with violence causing the death and injury to numerous persons. The turmoil in the valley had also resulted in the loss of thousands of lives or limbs and thousands of orphans. It is estimated that the conflict has caused around 100,000 as orphans (Mcgivering 2000). In addition, the health infrastructure was heavily damaged which was already inadequate further added to the miseries and thereby making people vulnerable to health problems (Amin and Khan 2009). Apart from it, the poor governance further led to malfunctioning of health sector in the valley. Such an environment led to the deterioration of overall health including the physical, mental and psychiatric health.

Due to the constant threat of violence and chaotic environment, the local people found themselves trapped in a multitude of stresses, disorders like anxiety and post-traumatic as well as high amounts of psychosocial problems. The mental health experts in the valley have pointed out that that there has been a tremendous increase in the number of stress and trauma related cases and the persistence of these psychological problems have in turn given rise to general health problems like cardiac diseases, diabetes and hypertension (Hassan and Shafi 2013). Although, the state of J and K established the Mental Health Act of 1987 to create a mechanism for monitoring and licensing health care facilities. But, it could not yield fruitful results as there are no recognised mental health care facilities in private sector. Though, such facility is only at government psychiatric hospital while such facilities are lacking at primary or secondary health care institutions in the Valley (The Hindu 2011).

Besides, no proper attention had been paid towards women's health as there is only one maternity hospital catering the whole population. However, the worst hit of the turmoil had been on the mental health of people. According to the report of Medicins and Frontiers-Holland, there had been a visible increase in the number of psychiatric problems in the valley (Medicins Sans Frontiers 2006). The number of patients visiting Srinagar's only psychiatric hospital was just 6 persons per day before the initiation of conflict. However, the number increased from 1990s – 2000 and reached 250-300 per day. While in 2003, the number of psychiatric patients further enhanced to 50,000 per annum and reached at the maxim in 2006 when 82000 patients visited the hospital (Rather 2013). The table 4.3 shows the mental health depression among various age groups in terms of percentage:

Sex Ratio	Age Group	Percentage
Male	15-25	64.61
Male	26-35	62.65
Female	15-25	68.66
Female	26-35	68.64
Overall	15-25	66.67
Overall	26-35	65.33

Table: 4.3. Mental health depression of various age groups in Kashmir Valley

Source: Amin and Khan, 2009.

The people in the valley have therefore largely suffered mainly in terms of mental health problems which have been neglected so far and also little had been done to modernize the present health care infrastructure. It is evident from the fact that the valley is having a single 100 bedded hospital for the psychiatric patients. In addition, there is also scarcity of mental health staff as there are only 8 psychiatrists in civil sector and no clinical psychologists (Amin and Khan 2009). Thus, the absence of basic health care infrastructure and proper services at primary and secondary level acts hindered the health concerns in the valley.

Education

The importance of education and its implication on the overall development is underpinned by many scholars in modern society. Proper education is essential for curbing population growth, eradicating poverty, achieving gender equality, ensuring sustainable development and maintaining peace and development. Even in the participation of people in decision making, educated masses are having edge over uneducated in the sense they would easily communicate with other participants. Also, there is a constitutional obligation to make compulsory and free education available to all children under the age group 06-14 years (Article 21(a) of the Indian constitution).

The official reports of J and K contended that the several hundred school buildings have been either burnt down or damaged in the 1990s due to the ongoing conflict in Kashmir valley. Since, 1989, as many as 891 educational buildings were destroyed in the course of military confrontation. The security threats, strikes and arrest of teachers have left the process of educational performance paralyzed in the valley (Rather 2013). Also, the dropout rate of the students tremendously increased due to the lack of safe and proper infrastructure of school buildings (The Tribune 2008). In addition, a student on an average day during the turmoil years had to face many hardships in his day today life like sightings of soldiers, military checkpoints and barracks located all around the valley (Habibullah 2008). All this had negative implications on the overall educational system of the valley.

The turmoil in the valley leads to frequent strikes (Hartaals) against the presence of Indian military which were often responsible for committing human rights violations. As a result, the people left their work, home and schools frequently to protest against the Indian occupation of their homeland (Arshad 2008). This process of shutting down brings everything to a standstill and had reverse consequences on the educational structure. Also, there is no contingency plan for the maintenance or promotion of educational institutions during the period of strikes which further brings the educational system on the back foot (Ibid). In addition, scores of widows of the conflict irrespective of their economic status give top priority to the education of their children. But, in order to support their household, they leave their schools early to earn the livelihood. Due to the reason, the students in the valley lead to drop out after secondary schooling (Parvaiz 2007). Though, the educational system in the valley suffered a lot due to the ongoing conflict vis-à-vis the poor governance on the part of state government.

Besides, the presence of security forces within and around the schools generates a sense of insecurity among the school going children. It is estimated that seventy nine percent of schools were at a distance of less than one kilometre from the nearest military bunker/camp (Ahmed 2013). Also, the security threats, strikes and arrest of teachers have left the process of educational performance paralyzed in the valley (Rather 2013).During the turmoil years, a student on an average day had to face many hardships in his day today life like sightings of soldiers, military checkpoints and barracks located all around the valley (Habibullah 2008). All this had negative implications on the overall educational system of the valley.

Thus, the valley of Kashmir had remained underdeveloped in terms of reduction of capital, employment and income, health, industrial sectors and so on. Also, due to the turmoil, the valley witnessed deaths, injuries, disappearances and restrictions on free movement making the life harder and more miserable. Further, the ongoing conflict had resulted in displacement of Kashmiri Pandits leading to the loss of human resources. In addition, there has been heavy damage to the infrastructure of the valley and the most important challenge had been the human rights violations. The above three features will be discussed and elaborated in the next section.

Impact of Conflict on the Kashmir Valley

Since the inception of militancy in the valley of Kashmir in early 1990s, it witnessed disturbances, turmoil and bloodshed from every nook and corner. Every individual remained in the state of constant fear and suspicion. Also, the valley remained one of the most militarised regions in the world operating under special laws like AFSPA and PSA. In addition, the conflict had created several challenges in terms of human loss, displacement of Kashmiri Pandits, infrastructural damage and more particularly the human rights violations.

Displacement of Kashmiri Pandits

The displacement of Kashmiri Pandits is among the unambiguous realities of the Kashmir conflict. Primarily, there are two dimensions related to the displacement of this community i.e., the internal turmoil and the external dimension. Internally, the violent militancy had erupted in the valley and externally, the presence of India-Pakistan hostilities of the conflict (Shekhawat 2006). However, most of the Pandits migrated from the valley claimed that they were forced to leave, though there are varying arguments about who compelled them to leave. Some claimed that the extremist armed groups forced them to do so, as there was an ultimatum from Hizbul Mujahideen published in a local newspaper on 14 April 1990 'Alsafa' that all Pandits should vacate from the valley within two days (Teng and Gadoo 1991). While, others claimed the circumstances prevailing in the Kashmir valley at that time forced them to leave as they were in minority. But, it is also the fact that the administration in the valley as well as the central government failed to provide any support to them and to assure their safety.

The displacement of Kashmiri Pandits from the valley is an integral part of the Kashmir conflict (Shekhawat 2006). It is estimated that about 2.5 to 3 Lakh people have been displaced from the valley since the violent conflict initiated. These displaced people have largely taken their shelter in Jammu region and Delhi. In Jammu, around 34,131 displaced families have been registered, while as in Delhi, there are around 19,338 registered families (Malhotra 2007). However, these families have been largely recognised as an important part of the valley in terms of as their role in generation of human capital resource as well as to influence household, individual and community behaviour. However, the migration of this community which changed their sociological and ecological setting had multiplicity of effects on their living conditions and survival (Arora and Dhingra 2005).

At the emotional and socio-cultural levels, there had been tremendous increase in family disruption, high level of stress mainly on the younger individuals and psychological disturbances among the displaced persons. It has been widely showed by various surveys conducted that the Kashmiri Pandit children are severely facing behavioural problems vis-a-vis depression, aggression, anxiety and sleeplessness (Ibid). The economy of the community has been also adversely affected leading to the great difficulties for their survival. In addition, most of them still live in temporary refugee camps which are overcrowded. The people living in these camps lack basic necessities and adequate facilities. Also, the Pandits residing in these camps lack employment and educational opportunities, and suffer from serious health issues like high occurrence of diseases, high death rates and psychological problems.

The displacement of Kashmiri Pandits has affected the valley in numerous ways. On the political level, their displacement jeopardized the secular ethos of Kashmir in which two religious communities interacted to each other from times immemorial. Culturally, the valley had endowed with the rich tradition of cultural diversity in which Hinduism, Sikhism and Islam had been living together for centuries. But, the conflict has deeply eroded the diversity of culture. Besides, the displacement led to the loss of human resources that might have been used to boost economy of the valley.

To sum up, over the two decades of the conflict in the valley, the displaced Kashmiri Pandits are still unable to return to their homeland. Despite, the state government's initiative to provide safe rehabilitation for them and assuring their safety, they are still feeling insecure to come back to the valley of Kashmir.

Infrastructural Damage

Conflict of any nature and of any kind has always a direct impact on the socio-economic structure of the society and the valley of Kashmir is being no exception to it. The ongoing conflict in valley has not only affected the development process, but has also led to heavy damage to the infrastructure. Due to the violent incidents, both private as well as public properties have been damaged, hundreds of educational buildings, government buildings, hospitals, private houses, bridges and shops have been gutted (Shekhawat 2009). The damaged and poor infrastructure however led towards low quality of life indicators and poor delivery of social justice. In contrary, the improved infrastructure would have increased productivity and competitiveness, so, thereby facilitating trade and foreign direct investment (Khan and Kiyawat 2013). Thus, the deficiencies in the infrastructure have reduced the valley's ability to participate in national trade activities which further reduced its economy.

The conflict in the valley have caused heavy damage to the infrastructure as it is estimated that from 1989 to 2002, more than 1,151 government buildings, 10729 private houses, 643 educational buildings, 1,953 shops, 337 bridges and 11 hospitals have been gutted due to this armed violence (Shekhawat 2009). Besides, the basic infrastructure in terms of communication system, roads, regular power supply and drinking water facilities also remained poor by the presence of violent environment.

The communication system in the valley remained inadequate and poor to serve the majority of population. This is evident from the fact that there is only one post office for an area of 60 sq. km as against the national average of 20 sq. km. The valley also faces problems in terms of cellular phones as the pre-paid connections lack the roaming facility outside the state. In addition, the short messaging services (sms) have been barred in the valley for a longer period of times. Besides, the internet facilities are often snatched by the government authorities, thereby making the people paralysed and disconnected with the rest of world. Similarly, as per the survey findings, the road length per 100 km in Kashmir is 35.71 km as against the national average of 104.64 km. This shows that the road density is very low as compared to other regions of India with large inter-district variations. Also, there are 2060 habitants in the valley which have no road connectivity at all (Shekhawat and Mahapatra 2008). Further, the valley lacked proper railway network and the majority areas remained un-benefitted to avail this service.

The valley is also facing acute electricity crises with regular 11-14 hour power cuts. In addition, there are around 25 Percent households in rural areas and about 2 Percent in urban areas with no electricity facilities (Ibid). However, the valley has the potential to generate 16, 480 megawatts of hydroelectric power implying that, it is boosted of having the total power generating capacity of nearly equivalent to the United States. If this energy resource is utilised to its full potential, it could become a major energy producer in the world. The electronic power generated would supply electricity to Pakistan, northern India and Central Asia (Jones. Et.al. 2010). By implication, the valley would be leveraged with huge revenue and generate employment.

In addition, the accessibility of safe drinking water is somehow lacking, indeed in rural areas, there seems to be huge shortage of safe drinking water. As a matter of fact, the safe drinking water is not accessible to 45 Percent of rural population while 4 Percent of urban population lack easy accessibility in the valley of Kashmir (Shekhawat and Mahapatra 2008). The lack of infrastructure e.g, road, electricity, railway in the valley of Kashmir pose serious challenge on the path of secure human life that would lead to flourish life. Infact, the gloomy picture itself revel more by suggesting that much could be done and achieved, if the conflict would not have been days of reality in valley of Kashmir.

Human Rights Violations

The valley of Kashmir faced the most prominent human security challenge in terms of human rights violations. The core human rights issues became an everyday occurrence in the valley mainly since the inception of armed conflict in late 1980's and early 1990's. Due to this violent conflict, thousands of civilians have died in regular clashes between separatist militants and security forces. The valley has mainly witnessed massive human rights violations after the New Delhi imposed the presidential rule in early 1990's. By the imposition of presidential rule, the Indian forces have been carrying on a brutal campaign against the Kashmiris in the valley who are fighting for their right of self-determination. The Indian forces have been granted by unrestricted arbitrary powers providing authority to them to use lethal force. Thus, the Indian government by giving legal sanctions to its forces leading to thousands of deaths, prisoners, persecutions and injuries to the people of Kashmir. Although, there is no state of emergency in the valley but still there are around 7 lack Indian forces stationed in the state mainly in the valley. The Indian government describes the situation as law and order problem (Mohiuddin 1997). If so, then how would the Indian government permit such laws to its citizens that seem if a rebellion is going on?

These special powers under the draconian laws like AFSPA and PSA, make Kashmiris to live in a state of constant fear of arbitrary arrests, enforced disappearance, sexual harassment, torture and custodial death. The nature of the powers conferred under the AFSPA reveals that there existed a condition of legal civil war against civilian population in the valley of Kashmir. Also, the act had sanctioned the existence of more than half a million armed forces and thereby making the valley of Kashmir as one of the most heavily militarised regions in the entire world. By giving special powers to the armed forces under the AFSPA, the Kashmiris regard their homeland not as a disturbed area but as an occupied zone and consider themselves as permanent prisoners of the state (Duschinshki 2009). Though, the act remained contentious from its very inception. Various observers have claimed that AFSPA is unconstitutional as it violates international humanitarian law. In addition, numerous civil society organisations have demanded the removal of the act and withdrawal of the troops from the valley as the act has been largely misused by the armed forces (Singh 2011).

Various reports have depicted the involvement of armed forces including the high ranked army officials in committing human rights violations in the valley. The Kashmiri based two human rights groups namely International People's Tribunal on Human Rights and Justice in Indian Administered Kashmir (IPTK) and Srinagar-based Association of Parents of Disappeared Persons (APDP) claimed that the top army officers have been involved in gross human rights violations. They prepared a report which revealed that hundreds of top ranked army personnel including 3 brigadiers of Indian Army, 3 lieutenant colonels, 78 majors, and 37 senior officials of federal Paramilitary forces and 25 captains were offenders of several crimes and human rights abuses in the valley. The report also showed that these perpetrators have not been prosecuted in the state of Jammu and Kashmir (The Hindu 2012).

In addition, according to the same report made by IPTK and APDP published on international business times (2012), fair investigations have not been undertaken by India and its courts, besides the availability of evidences of human rights abuses. The report identified 500 individual human rights offenders by reviewing 214 cases which included 235 army officers, 123 military personnel, 111 Jammu and Kashmir personnel and around 31 state backed militants. The report further said that an estimated of 8,000 persons had disappeared in the past 20 years vis-à-vis causing to the death of over 70,000 people. Also, it cited disclosures of more than 6,000 unmarked, unknown and mass graves.

The most notable forms of human rights violations occurring in the valley include arbitrary arrests, extra-judicial killings, disappearances, tortures and rapes. These violations committed by the Indian forces have been well documented by several Human Rights Organisations such as Amnesty international, Asia watch, People's Union for Democratic Rights (PUDR), People's Union for Civil Liberties (PUCL) and local Human Rights Groups like Jammu and Kashmir Human Rights Commission (JKHRC) and People's Basic Rights (protection) Committee (PBRC) (Ibid).

(a) Custodial/Extrajudicial killings

The direct impact of the on-going conflict in the valley of Kashmir has been a large number of custodial/extrajudicial killings either by security personnel or by paramilitary forces. Most of these killings have resulted from firing on peaceful possessions, encounter killings, funeral processions, killings during search operations and cordon and in cross-firing between the militants and paramilitary forces. From 1989 onwards, the valley of Kashmir had witnessed over more than 119,789 arrests of civilians. Out of these arrests made by armed forces, 6989 civilians have been killed in the custodians (Tabassum 2012). Apart from the custodial killings, the valley of Kashmir has seen multiple deaths of civilians during the recent protests in 2008, 2010, 2013 and 2016. In the year 2008, the civil unrest broke out in the valley against the human rights violations committed by the armed personals leading to the death of 63 civilians. Similar unrest again visited the valley of Kashmir in 2010 on the issue of land transfer to shrine board. The majority of the people in the valley disliked and unflavoured the move of the government resulting in clashes between the civilians and the armed forces. The agitation which lasted for more than three months causing massive human rights violations leading to the death of more than 200 civilians. In addition, the valley further witnessed almost 20 civilian killings by the police, paramilitary and armed personals. Recently, in 2016, more than 90 civilians have lost their lives, over 15, 000 people injured and more than 12, 000 were arrested and detained.

(b) Enforced Disappearance

The most traumatic consequence of the conflict in the valley has been missing of people in detention. Under international human rights law, enforced disappearances are those committed by the state by various security agencies, troops and police and are regarded as crimes against humanity. The number of enforced disappearances went on increasing since early 1990's as the police in the valley do not entertain missing reports of those persons. According to the report of International People's

Tribunal on Human Rights Violations in Kashmir (2010), the number of people who have disappeared in the valley since 1989 ranging from 8,000-10,000. The report also revealed that most of them have been killed either in custody of the security personnel or the army.

(c) Torture

The torture in the valley of Kashmir seems to be operationalized by the Indian forces namely the army, BSF, and CRPF as a basic tool and policy to combat with Kashmiri's, since the late 1980's. The several human rights groups showed that everyone including children, women, men, elderly and sick were subjected to worst forms of cruel, degrading and inhuman punishment or treatment. Also, the methods adopted during torture were extremely humiliating and cruel. The most common forms of torture with which the people of valley were dealt include both psychological and physical.

Physical Torture: According to the Amnesty International Report (1991), the practices of physical torture mainly adopted by the Indian forces in the Kashmir Valley include:

- Prolonged and severe beating as well as giving electric shocks to sensitive parts of the body.
- Burning the skin with hot irons and pouring the body with hot water.
- Rolling the wooden log or heavy metal across the body in order crush the joints and bones from legs up to the neck.
- Putting chilli powder in the eyes.
- Molestation and raping of Kashmiri women and girls.

Psychological Torture: Apart from imparting physical pain and damaging the body parts, the Indian forces had cost heavily upon the psychological health of the people of valley. Such methods were adopted during the torture which impacts upon the psychological health of the people as:

- Compelling the victims to eat human excrement.
- Inflicting torture on women and children in front of the family members.
- Enactment of fake executions.
- Using of rude and abusing language (Report of Asia Watch and Physicians for Human Rights 1992).

(d) Rape Cases

The greatest atrocious and dreadful crime in terms of humanity is the rape. It has been used as a tool of humiliating an individual as well as community in destroying their honour. In the valley of Kashmir, rape by Indian forces since early 1990 has become more frequent. This inhuman treatment mostly occurred during cordon, search operations and crackdowns conducted by armed forces, when men are held for identification in schoolyards or parks. During these search operations, the security forces enter into the homes and rape the Kashmiri women (Ibid). The most unbearable among the rape cases in the valley include the Kunan Poshpora case in Kupwara district in which at least 23 women and as many as one hundred girls were repeatedly raped by the Indian army in the homes at gun point (Ghosh 2013). The victims included all age groups, as the youngest was a girl of 13 years and the oldest victim was of 80 years. The table 4.4. on the next page shows impact of violence on the valley of Kashmir from 1989-2012 in terms of human rights violations:

Total killings	93,716
Civilians arrested	119,789
Custodial killings	6,989
Children orphaned	107,436
Women widowed	22,763
Women molested/gang raped	10,350
Structures destroyed	105,936

Table: 4.4. Impact of violence on the valley in terms of human rights violations

Source: Tabassum, 2012

However, no actions have been taken till the date against the offenders despite the availability of evidences and various reports in the journals and newspapers and also by several NGO groups both national as well as international (Report of International People's Tribunal on Human Rights Violations in Kashmir 2010).

Looking back to the human security in Kashmir valley, no one would claim that the prospect and the condition of human life are under their control. Indeed, numerous challenges and hardships are imposed upon human life; which are beyond figment of our imagination. The key variables of human security namely education and poverty are having the worse record. For instance, the poverty in the valley has not been somehow able to improve; similarly, the education, even though one would claim crude measurement, namely Gross Enrolment Ratio (GER) is not as good as other states of India. Besides, the on-going conflict and militarism has many times destroyed the school buildings or for months schools are closed. Besides, the major human development indicators (e.g. namely poverty or unemployment) are not encouraging either, and number of factors would be operating. On poverty level, the valley has large number of people in BPL category in comparison to national average as well as to

Jammu region. Similarly, the unemployment in the valley has somehow, been caused by the on-going conflict and disruption.

The geographical location of valley gives edge over other area in term of developing tourism industries, but the conflict and its consequences did not allow developing this sector. Needless to mention, tourism is, usually, understood to be major sector in which large number of employment is being created. But the educated unemployed could not get job due to lack of job prospect. Further, the inadequate infrastructure, e.g. road and so on, has partly accounted for non-development of market. Hence, the economic prospect of state is not as good as it might be supposed.

The major indicator of human security is sense of safety and political liberty, but the on-going conflict and militarism has made it hard by unnecessary prison, torture in jail, sexual assaults and many times the rape of women by military forces has become the common feature of the valley, especially after 1990s. As many reports and daily newspapers make it clear that the conflict has been resulted in psychological trauma and pain due to losses of near and dear in the region, even many children has been living in orphan; thus, the life and its surrounded fear in the valley is paramount and culminated in curtailing the political freedom, e.g., right to protest. Hence it would not be exaggeration to claim that multiplicity of underdevelopment of valley is caused by on-going conflict and heavy imposition of many acts add the wounded the valley.

Throughout the subcontinent's postcolonial history, Indian rule in Kashmir has included militarization, repression, economic deprivation, and indiscriminate violence, including at various times, the denial of democratic processes, the manipulation of elections, and the jailing of political leaders most notably Sheikh Abdullah - who dominated regional politics for over thirty years until his death in 1982 (Behera 2000, Bose 2003). Also, the patterns of sustained disenfranchisement and marginalization have entrenched Kashmiri collective feelings of alienation from the Indian state. Kashmiri's today use the language of self-determination as a way of demanding an opportunity to express their collective will in relation to their own political future.

CHAPTER – 5
CONCLUSION

The concept of security has undergone drastic changes since the end of cold war. The focus has largely shifted from territorial defence to people's or community protection. It is within this context that human security emerged in the contemporary era and achieved a top priority concern. It is more human centric approach and included within its ambit the issues of social justice, economic development, human rights issues, environmental protection, disarmament and so on. However, there is no single unified definition to the concept. To UNDP, human security evolved out of the threats from economic, food, health, environmental, personal, community and political security. Similarly, the concept has been differently interpreted by the Canadian and Japanese governments. The former relates it with the protection of human beings from violent threats in complex situations e.g., war, emergencies etc., while the latter focuses on the promotion of fulfilment of human development as well as human needs.

On the other hand, the critics have challenged it on its inclusiveness including an ample variety of conceptual frameworks which makes it meaningless as an analytical tool. Also, the notion of human security have been challenged by many on the basis of allotting prioritization to the threats induced in the concept, thereby making it difficult to be implemented and practised. Thus, besides being so ambiguous in conceptual clarification, the ultimate aim of the concept is centred on the individual or community rather than on states.

Apart from all the criticisms labelled against it, it is apparent that the concept of human security encompasses global issue in the sense that the threats are equally common to all countries. Although, the intensity of threats may vary from country to country, but the fact is that they are continuously emerging and are interrelated to each other. Accordingly, if menaces of human security are visualised somewhere in the world, it means all nations would be affected. For instance, environmental degradation, terrorism, drug trafficking, social disintegration etc. are no longer national or isolated events but their consequences can affect or reach out at global level. Thus, it is the collective responsibility of all nations-states to evolve around flourishment of human being. In this way, the valley of Kashmir gives heavy exercise to see what it offers. Also, the issues pertaining to the valley of Kashmir provides ample scope to study the challenges of human security.

While, looking at the Kashmir valley, there are several human security challenges confronting the people, mainly due to the internal conflict as well as India-Pakistan rivalry. The conflict initiated mostly after the British India's division in 1947 into two dominions of India and Pakistan. Both the dominions wished to integrate the valley to their own side in order to broaden their political dimensions. The Indian leaders viewed that it is fruitful to merge Kashmir to its territorial ambit as its claim of secularist principle would be enhanced. Pakistan on the other hand claimed that majority of the population of valley belongs to Muslims, along with bearing close proximity in terms of geographical connectivity and socio-cultural set up with Pakistan. Under such circumstances, the valley of Kashmir emerged as a disputed territory between the two countries on which three wars have been fought. The dispute was ultimately referred to United Nations Security Council in 1948 by the Indian government, where decision was reached to conduct plebiscite under the UN supervision. But sooner the Indian government developed cold feet on the matter due to internal political pressures and kept that decision aside. As a result, no plebiscite has taken place so far and the reason behind this can be understood as the reluctance as well as unwillingness imparted on the part of Indian government.

In the internal context, the policies adopted by the Indian government towards the Jammu and Kashmir state played a role in escalating the conflict of the Kashmir valley by greatly interfering into the state matters and in abolishing the special provisions allotted to it. Also, the setting up of puppet governments through undemocratic means, subsequently imposing presidential rule, eroding the autonomy granted under article 370, opportunistic accords and alliances with the national conference, electoral malpractices and using of armed forces to suppress agitation had greatly led towards escalating the Kashmir conflict. Further, the heavy militarization prevailing in the valley has been responsible for human rights violations vis-à-vis have undermined the ethos of democracy.

Apart from the internal dimensions, the valley of Kashmir suffered several challenges in terms of economic, political and socio-cultural factors. The valley witnessed acute problems of poverty as is evident that almost 13 lakh people are living under the Below Poverty Line. The official estimates showed that every fifth person is falling under the BPL. Also, the districts affected with militancy as well as militarization recorded the highest rate of poverty. The other human security challenge confronting the people of valley is the problem of unemployment. The absence of desired industrial development and the lack of absorption in the private sector have resulted into alarming unemployment problems. Presently, the official records indicate that there are about 3, 21,526 unemployed youth in the valley, besides lagging behind in terms of literacy rate, having 30th position at national level.

Moreover, the agricultural sector besides being the major source of livelihood remained poor both on the part of state's negligence as well as geographical location. The valley witnessed under development in agriculture mostly due to the lack of irrigational facilities and deficiency in agricultural research. On the other side, the existence of distant markets for export as well as unorganised market facilities became a stumbling block in hindering the advancement of this sector. Moreover, the industrial sector having a great potential of employment generation and mitigating poverty, remained under developed or untouched. In absence of any heavy industry, the small industrial sector e.g., Handicrafts and Tourism which were the traditional income source of a large chunk of population, also suffered heavily due to the conflict and were unable to fulfil the needs, and thereby adding more complexities and challenges for survival to the people of the valley.

The impact of conflict not only cost upon the employment and industry, but had also badly affected the public health. It had deteriorated the physical, mental and psychiatric health, along with causing death and injury to several thousand people. In addition, the already inadequate public health infrastructure was heavily damaged, which further added to the miseries of the people and thereby making them vulnerable to health problems. Moreover, the valley showed poor output in terms of educational progress and advancement. The persistent violent environment has paralysed the educational system and the outcome has been unsustainable development and chaos among the people.

In early 1990's, the valley witnessed the displacement of second largest population community commonly known as 'Kashmiri Pandits', leading to the loss of human resource. However, there exists a dilemma regarding the factors responsible behind their displacement. The question remained unanswered whether the initiation of violent environment had generated a feeling of insecurity in them, or the inception of militant groups prevailing in the valley forced them to leave. But, it is a fact that both the state as well as central governments failed to provide them any substantial support, along with assuring their safety and security. Whatsoever, being the reasons, the displacement has badly affected the valley as it had endangered the secular ethos of valley. Also, the rich heritage of cultural diversity and harmony has been eroded, besides leading to the economic loss.

The other dimensions of the conflict have been the heavy damage to the infrastructure leading towards low quality of life indicators. Thousands of government buildings, private houses, educational buildings, shops, bridges and hospitals have been gutted due to the violent environment in the valley. Also, the basic infrastructure in terms of communication system, roads, regular power supply and drinking water facilities remained inadequate to fulfil the daily requirements of the people of valley. In addition, the deficiencies in the infrastructure have reduced the valley's ability to participate in national trade activities which further reduced its economy. The closer proximity with international border allows the opportunities of international trade with neighboring countries that might have been boosted local economy, if the infrastructure were adequate.

Besides, the valley of Kashmir is one of the world's heavily militarized regions in which the two nations namely India and Pakistan are at stake. This region is also mentioned by many as one of the most dangerous places on earth. The presence of such a huge military has generated a sense of alienation as well as a feeling of insecurity among the people in the valley. It seems that they are caged in iron wall of militarization and severe limitation has been imposed upon human life, these are not figments of our imagination. As a matter of fact, the militarization has been explicitly (or implicitly) involved in gross violation of human rights that ranges from rapes of minor girls to taking away of human lives. Needless to mention that the draconian colonial laws like AFSPA and PSA; which guarantees and assures the armed forces to have a free hand in committing human rights violations are operational till the date. Surprisingly enough, these are the state of affair under democracy.

The multi-facet challenges of Kashmir Valley that goes to the core of political realm, make human security conception somehow obsolete in the term of state vis-a-vis society at large in which one is overpowered over other. Also, how the flourishment of person could be evaluated in the light of these realities. Since any exercise of evaluation must be difficult task especially, if that rest with some of reality in which the shadow of state apparatus is visible. Hence, it is in deeper trap to gauges the process through which people's life unfold in term of opportunity (or flourishment); more so, if one takes the present scenario of human development in valley of Kashmir, it is easy to testify, e.g. the presence of heavy number of unemployed (or underemployed) youth. Needless to mention, the political freedom demands to nurture the countervailing power of other institution of society of which one could be civil society, yet the beneath is hollowed by the constant human rights violations and so on. Thus, the paradox of the applicability of human security in the valley of Kashmir become apparent; but the ray of hope lie in the 'dialogue' which is somehow missing between all the three party involved - namely India, Pakistan and the people of Valley.

Since the state of human security prevailing in the valley is far from commendable, it is necessary to look into alternative dimensions by throwing light on some of the vital perspectives. The study provides an ample ground to consider these from human security dimensions. Firstly, the United Nations could have subverted the huge catastrophe that Kashmir went through escalating of tension between India and Pakistan in last more than six decades. That shows how minimum role the UN has played in resolving the tension, given the fact, one of its core principles is to mitigate (or eradicate) tension, so that peace could avail. Yet, the on-going tension in Kashmir valley demand the United Nation to take pro-active role. Secondly, both the involved countries namely India and Pakistan have been using confidence building measures (CBM'S) as a means to resolve the conflict, that would, otherwise, have been used as end in itself. It reflects political blindness on the path of resolving the tension on both side of the table. Indeed, political unwillingness essentially lie in not looking from vantage point of people of Kashmir valley about which human security conception evolved around.

Also, from perspective of India, it is neither desirable nor possible to go back to golden age of Kashmir valley in which it had independence. There are mainly two reasons which cannot be under-estimated. One is that, it is against the general public opinion as well as India's desire to maintain its secular ethos that valley of Kashmir has been seen upon. Secondly, the Indian union was very desperate in integrating the valley within its territorial ambit since very beginning, as it can be seen on the ground of massive budget outlay for the valley. Hence, the greater autonomy or restoration of Article 370 would be one of the finest solution in which the slogan of wedding the bloody shirt will be evaporated. Similarly, the sluggish economy of the valley of Kashmir can be improved if infrastructure develops, along with the development of tourism and other industries. Also, the geographical proximity of valley with other nation state gives opportunities to have trade with them, where the valley has comparative advantage.

Above all, the militarization of valley has resulted in massive human rights violations, but the time has come to roll back the process of oppression and exploitation. As any human life can only flourish in freedom, but the shadow of militarization in the valley of Kashmir has casted its print immensely to curtail the freedom. That would not be anyway claimed as it is assisting in flourishment of human life. Hence to any solution, first and foremost lies in demilitarization. As most of the solutions offered so far failed to look beyond the interests of both the nations involved, the Human security approach brings a fresh air to the conflict prone region by relocating the security from people's perspective. In this context the human security approach can be used as a tool to understand and offer solutions to the issues of militarised regions across the world in general and the valley of Kashmir in particular.

Appendix 1

Treaty of Amritsar

March 16, 1846

(Courtesy: Department of Law, Justice & Parliamentary Affairs, Government of Jammu and Kashmir)

The treaty between the British Government on the one part and Maharajah Gulab Singh of Jammu on the other concluded on the part of the British Government by Frederick Currie, Esq. and Brevet-Major Henry Montgomery Lawrence, acting under the orders of the Rt. Hon. Sir Henry Hardinge, G.C.B., one of her Britannic Majesty's most Honourable Privy Council, Governor-General of the possessions of the East India Company, to direct and control all the affairs in the East Indies and by Maharajah Gulab Singh in person - 1846.

Article 1

The British Government transfers and makes over for ever in independent possession to Maharajah Gulab Singh and the heirs male of his body all the hilly or mountainous country with its dependencies situated to the eastward of the River Indus and the westward of the River Ravi including Chamba and excluding Lahol, being part of the territories ceded to the British Government by the Lahore State according to the provisions of Article IV of the Treaty of Lahore, dated 9th March, 1846.

Article 2

The eastern boundary of the tract transferred by the foregoing article to Maharajah Gulab Singh shall be laid down by the Commissioners appointed by the British Government and Maharajah Gulab Singh respectively for that purpose and shall be defined in a separate engagement after survey.

Article 3

In consideration of the transfer made to him and his heirs by the provisions of the foregoing article Maharajah Gulab Singh will pay to the British Government the sum of seventy-five lakhs of rupees (Nanukshahee), fifty lakhs to be paid on or before the 1st October of the current year, A.D., 1846.

Article 4

The limits of territories of Maharajah Gulab Singh shall not be at any time changed without concurrence of the British Government.

Article 5

Maharajah Gulab Singh will refer to the arbitration of the British Government any disputes or question that may arise between himself and the Government of Lahore or any other neighboring State, and will abide by the decision of the British Government.

Article 6

Maharajah Gulab Singh engages for himself and heirs to join, with the whole of his Military Forces, the British troops when employed within the hills or in the territories adjoining his possessions.

Article 7

Maharajah Gulab Singh engages never to take to retain in his service any British subject nor the subject of any European or American State without the consent of the British Government.

Article 8

Maharajah Gulab Singh engages to respect in regard to the territory transferred to him, the provisions of Articles V, VI and VII of the separate Engagement between the British Government and the Lahore Durbar, dated 11th March, 1846.

Article 9

The British Government will give its aid to Maharajah Gulab Singh in protecting his territories from external enemies.

Article 10

Maharajah Gulab Singh acknowledges the supremacy of the British Government and will in token of such supremacy present annually to the British Government one horse, twelve shawl goats of approved breed (six male and six female) and three pairs of Cashmere shawls.

This Treaty of ten articles has been this day settled by Frederick Currie, Esq. and Brever-Major Henry Montgomery Lawrence, acting under directions of the Rt. Hon. Sir Henry Hardinge, Governor-General, on the part of the British Government and by Maharajah Gulab

Singh in person and the said Treaty has been this day ratified by the seal of the Rt. Hon. Sir

Henry Hardinge, Governor-General. Done at Amritsar the sixteenth day of March, in the year of our Lord one thousand eight hundred and forty-six, corresponding with the seventeenth day of Rubee-ul-Awal (1262 Hijri).

Sd/-
H. Hardinge (Seal)
F. Currie
H. M. Lawrence

Appendix 2

Instrument of Accession of Jammu and Kashmir State

(Courtesy: Department of Law, Justice & Parliamentary Affairs, Government of Jammu and Kashmir)

Whereas the Indian Independence Act, 1947, provides that as from the fifteenth day of August, 1947, there shall be set up an independent Dominion known as INDIA, and that the Government of India Act 1935, shall with such omissions, additions, adaptations and modifications as the Governor General may by order specify, be applicable to the Dominion of India.

AND WHEREAS the Government of India Act, 1935, as so adapted by the Governor General, provides that an Indian State may accede to the Dominion of India by an Instrument of Accession executed by the Ruler thereof.

Now, Therefore

I, Shriman Inder Mahinder Rajrajeswar Maharajadhiraj Shri Hari Singhji, Jammu & Kashmir Naresh Tatha Tibbet adi Deshadhipati, Ruler of Jammu & Kashmir State, in the exercise of my Sovereignty in and over my said State do hereby execute this my Instrument of Accession and

1. I hereby declare that I accede to the Dominion of India with the intent that the Governor General of India, the Dominion Legislature, the Federal Court and any other Dominion authority established for the purposes of the Dominion shall by virtue of this my Instrument of Accession but subject always to the terms thereof, and for the purposes only of the Dominion, exercise in relation to the State of Jammu & Kashmir (hereinafter referred to as "this State") such functions as may be vested in them by or under the Government of India Act, 1935, as in force in the Dominion of India, on the 15th day of August 1947, (which Act as so in force is hereafter referred to as "the Act').

2. I hereby assume the obligation of ensuring that due effect is given to provisions of the Act within this State so far as they are applicable therein by virtue of this my Instrument of Accession.

3. I accept the matters specified in the schedule hereto as the matters with respect to which the Dominion Legislature may make law for this State.

4. I hereby declare that I accede to the Dominion of India on the assurance that if an agreement is made between the Governor General and the Ruler of this State whereby any functions in relation to the administration in this State of any law of the Dominion Legislature shall be exercised by the Ruler of the State, then any such agreement shall be construed and have effect accordingly.

5. The terms of this my Instrument of Accession shall not be varied by any amendment of the Act or the Indian Independence Act, 1947, unless such amendment is accepted by me by Instrument supplementary to this Instrument.

6. Nothing in this Instrument shall empower the Dominion Legislature to make any law for this State authorizing the compulsory acquisition of land for any purpose, but I hereby undertake that should the Dominion for the purpose of a Dominion law which applies in this State deem it necessary to acquire any land, I will at their request acquire the land at their expense, or, if the land belongs to me transfer it to them on such terms as may be agreed or, in default of agreement, determined by an arbitrator to be appointed by the Chief Justice of India.

7. Nothing in this Instrument shall be deemed to commit in any way to acceptance of any future constitution of India or to fetter my discretion to enter into agreement with the Government of India under any such future constitution.

8. Nothing in this Instrument affects the continuance of my Sovereignty in and over this State, or, save as provided by or under this Instrument, the exercise of any powers, authority and rights now enjoyed by me as Ruler of this State or the validity of any law at present in force in this State.

9. I hereby declare that I execute this instrument on behalf of this State and that any reference in this Instrument to me or to the Ruler of the State is to be construed as including a reference to my heirs and successors.

Given under my hand this 26th day of October, nineteen hundred and forty seven.

Sd/-
Hari Singh
Maharaja Dhiraj of Jammu and
Kashmir State.

(Acceptance of Instrument of Accession)

I do hereby accept this Instrument of Accession.

Dated this twenty seventh day of October,

Nineteen hundred and forty seven.

Sd/-

Mountabatten of Burma

Governor General of India

27th October 1947

Appendix 3

THE ARMED FORCES (JAMMU AND KASHMIR) SPECIAL POWERS ACT, 1990

No. 21 of 1990

[10th September, 1990.]

An Act to enable certain special powers to be conferred upon members of the armed forces in the disturbed areas in the State of Jammu and Kashmir.

BE it enacted by Parliament in the Forty-first Year of the Republic of India as follows:

1. Short title, extent and commencement.

(1) This Act may be called the Armed Forces (Jammu and Kashmir) Special Powers Act, 1990.

(2) It extends to the whole of the State of Jammu and Kashmir.

(3) It shall be deemed to have come into force on the 5th day of July, 1990.

2. Definitions. In this Act, unless the context otherwise requires,—

(a) “armed forces” means the military forces and the air forces operating as land forces and includes any other armed forces of the Union so operating

(b) “disturbed area” means an area which is for the time being declared by notification under section 3 to be a disturbed area;

(c) all other words and expressions ‘used herein, but not defined and defined in the Air Force Act, 1950, or the Army Act, 1950,2 shall have the meanings respectively assigned to them in those Acts.

3. Power to declare areas to be disturbed areas.

If, in relation to the State of Jammu and Kashmir, the Governor of that State or the Central Government, is of opinion that the whole or any part of the State is in such a disturbed and dangerous condition that the use of armed forces in aid of the civil power is necessary to prevent—

(a) activities involving terrorist acts directed towards overawing the Government as by law established or striking terror in the people or any section of the people or alienating any section of the people or adversely affecting the harmony amongst different sections of the people;

(b) activities directed towards disclaiming, questioning or disrupting the sovereignty and territorial integrity of India or bringing about cession of a part of the territory of India or secession of a part of the territory of India front the Union or causing insult to the Indian National Flag, the Indian National Anthem and the Constitution of India, the Governor of the State or the Central Government, may, by notification in the Official Gazette, declare the whole or any part of the State to be a disturbed area.

Explanation.—In this section, "terrorist act" has the same meaning as in Explanation to article 248 of the Constitution of India as applicable to the State of Jammu and Kashmir.

4. Special powers of the armed forces.

Any commissioned officer, warrant officer, non-commissioned officer or any other person of equivalent rank in the armed forces may, in a disturbed area,—

(a) if he is of opinion that it is necessary so to do for the maintenance of public order, after giving such due warning as he may consider necessary, fire upon or otherwise use force, even to the causing of death, against any person who is acting in contravention of any law or order for the time being in force in the disturbed area prohibiting the assembly of five or more persons or the carrying of weapons or of things capable of being used as weapons or of firearms, ammunition or explosive substances;

(b) if he is of opinion that it is necessary so to do, destroy any arms dump, prepared or fortified position or shelter from which armed attacks are made or are likely to be made or are attempted to be made, or any structure used as training camp for armed volunteers or utilized as a hide-out by armed gangs or absconders wanted for any offence;

(c) arrest, without warrant, any persons who has committed a cognizable offence or against whom a reasonable suspicion exists that he has committed or is about to commit a cognizable offence and may use such force as may be necessary to effect the arrest;

(d) enter and search, without warrant, any premises to make any such arrest as aforesaid or to recover any person believed to be wrongful restrained or confined or any property reasonably suspected to be stolen property or any arms, ammunition or explosive substances believed to be unlawful kept in such premises, and may for that purpose use such force as may be necessary, and seize any such property, arms, ammunition or explosive substances;

(e) stop, search and seize any vehicle or vessel reasonably suspected to be carrying any person who is a proclaimed offender, or any persons who has committed a noncognizable offence, or against whom a reasonable suspicion exists that he has committed or is about to commit a noncognizable offence, or any person who is carrying any arms, ammunition or explosive substance believed to be unlawfully held by him, and may, for that purpose, use such force as may be necessary to effect such stoppage, search or seizure, as the case may be.

5. Power of search to include powers to break open locks, etc.

Every person making a search under this Act shall have the power to break open the lock of any door, almirah, safe, box, cupboard, drawer, package or other thing, if the key thereof is withheld.

6. Arrested persons and seized property to be made over to the police.

Any person arrested and taken into custody under this Act and every property, arms, ammunition or explosive substance or any vehicle or vessel seized under this Act, shall be made over to the officer-incharge of the nearest police station with the least possible delay, together with a report of the circumstances occasioning the arrest, or as the case may be, occasioning the seizure of such property, arms, ammunition or explosive substance or any vehicle or vessel, as the case may be.

7. Protection of persons acting in good faith under this Act.

No prosecution, suit or other legal proceeding shall be instituted, except with the previous sanction of the Central Government, against any person in respect of anything done or purported to be done in exercise of the powers conferred by this Act.

8. Repeal and saving.

(1) The Armed Forces (Jammu and Kashmir) Special Powers Ordinance, 1990, is hereby repealed.

(2) Notwithstanding such repeal, anything done or any action taken under the said Ordinance shall be deemed to have been done or taken under the corresponding provisions of this Act.

V.S. RAMA DEVI,
Secy. to the Govt. of India

GOVERNMENT OF JAMMU AND KASHMIR,

CIVIL SECRETARIAT—HOME DEPARTMENT.

SRO NO. SW 4

Dated 6-7, 1990

In exercise of the powers conferred under section 3 of the Armed Forces (Jammu and Kashmir) Special Powers Ordinance, 1990, the Governor of Jammu and Kashmir hereby notifies the areas given in the Schedule to this notification as Disturbed Areas.

(Sd.)
Additional Chief Secretary (Home),
Jammu and Kashmir Government.

SCHEDULE

1. Areas falling within 20 Kms. of the Line of Control in the Districts of Rajouri and Poonch.

2. Districts of Anangtnag, Baramulla, Badgam, Kupwara, Pulwama and Srinagar.

(Sd.)

Additional Chief Secretary (Home),
Jammu and Kashmir Government

Government of Jammu and Kashmir
Civil Secretariat Home Department
NOTIFICATION
SRINAGAR, THE 10TH AUSUGT, 2001

SRO 351: Whereas the Governor is of the opinion that the State is in such a disturbed condition that the use of Armed Forces in the aid of civil power is necessary to prevent the activities involving terrorists acts directed towards striking terror in the people;

Now, therefore, in exercise of the powers conferred by section 3 of the Armed Forces (Jammu and Kashmir) Special Powers Act, 1990, the Governor hereby declares the districts of Jammu, Kathu, Udhampur, Poonch, Rajouri and Doda to be disturbed areas in addition to districts, Srinagar, Budgam, Anantnag, Pulwama, Baramulla and Kupwara which stand already so declared.

Appendix 4

Jammu and Kashmir Public Safety Act, 1978

Act No. 6 of 1978,

CHAPTER I

PRELIMINARY

1. Short title and extent.

(1) This Act may be called the Jammu and Kashmir Public Safety Act, 1978.

(2) It extends to the whole of Jammu and Kashmir State

2. Definitions.

In this Act, unless there is anything repugnant in the subject or context, -

(1) "the Code"' means the Code of Criminal Procedure, Smvat 1989;

(2) "notified" and "notification"' mean notified and notification respectively in the Government Gazette.

CHAPTER II

ACCESS TO CERTAIN PREMISES AND AREAS

3. Prohibited places.

(1) If as respects any place the Government considers it necessary or expedient that special precautions should be taken to prevent the entry of unauthorised persons, the Government may, by notified order, declare that place to be a prohibited place.

(2) No person shall, without the permission of the Government or the authority specified by the Government, enter, or be on or in, or pass over, or loiter in the vicinity of, any prohibited place.

(3) Where in pursuance of sub-section (2) any person is granted permission to enter, or to be on or in, or to pass over, a prohibited place, that person shall, while acting under such permission comply with such orders for regulating his conduct as may he given by the Government or the authority specified by the Government.

(4) Any Police Officer, or any other person authorised in this behalf by the Government, may search any person entering or seeking to enter or being on or in, or leaving a prohibited place, and any vehicle, aircraft or article brought in by such person, and may, for the purpose of the search, detain such person, vehicle, aircraft and article.

Provided that no female shall be searched in pursuance of this sub-section except by a female.

(5)If any person is in a prohibited place in contravention of this section, then, without prejudice to any other proceedings which may be taken against him, he may be removed therefrom by any Police Officer not below the rank of a Sub Inspector or by any other person authorised in this behalf by the Government.

(6)If any person is in a prohibited place in contravention of any of the provisions of this section he shall be punishable with imprisonment for a term which may extend to one month, or with fine, or with both.

4. Protected areas.

(1)If the Government considers it necessary or expedient in the interests of the defence or security of the State to regulate the entry of person into any area, it may, by a notified order declare the area to be a protected area, and thereupon, for so long as the order is in force, such area shall be protected area for the purpose of this Act.

(2)The Government or the authority specified by the Government may regulate the entry of any person into a protected area.

(3)If any person is in a protected area in contravention of the provisions of any order passed under thin section then, without prejudice to any other proceedings which may be taken against him, he may be removed therefrom by or under the direction of any police officer not below the rank of a Sub-Inspector.

(4)If any person is in a protected area in contravention of any of provisions of this section, he shall be punishable with imprisonment for a term which may extend to two months, or with fine, or with both.

5. Forcing or evading a guard.

Any person who effects or attempts to effect entry into a prohibited place or a protected area after taking precautions to conceal his entry or attempted entry from any person posted for the purpose of protecting or of preventing or controlling access to such place or area shall be punishable with imprisonment for term which may extend to three months, or with fine, or with both.

CHAPTER III MAINTENANCE OF COMMUNAL AND REGIONAL HARMONY

6. Power to prohibit circulation within the State or entry into the State of certain documents.

(1)The Government, or any authority authorised by it in this behalf, if satisfied that such action is necessary for the purposes of preventing or combating any activity prejudicial to the maintenance `of communal or sectarian, or regional harmony effecting or likely to affect public order, may, by notified order, regulate or restrict the circulation within the State, or prohibit or restrict the importation into the State, of any document:

Provided that no such order shall remain in force for more than three months from the making thereof unless before the expiry of such period, and in case the High Court does not otherwise direct, the Government, by an order made in the like manner, extend it by any period not exceeding three months at a time as it thinks fit, so however, that the total period of the original order does not exceed one year

Provided further that a person aggrieved by such order may, within ten days of the passing thereof, make a representation to the Government which may, on consideration, confirm, modify or rescind the order within 21 days of the making of the representations, after giving the aggrieved party an opportunity of being heard

Provided also that in case the representation is rejected by the Government, the aggrieved person may within a period of two months from the date of the order rejecting the representation, apply to the High Court to set aside the order. Every such application shall be heard and determined by a Special Bench of the High Court composed of three Judges.

(2)Any person who contravenes an order made under this section shall be punishable with imprisonment for a term which may extend to three months or with fine or with both.

(3)In the event of disobedience of an order made under sub-section (1), the Government or the authority issuing the order, may without prejudice to the penalty to which the person guilty of the disobedience is liable under sub-section (2), order the seizure of all copies of, any such document.

7. Removal of doubts.

For the removal of doubts it is hereby declared that the restrictions imposed by section 6 on the rights conferred by clause (1) of Article 19 of the Constitution of India shall he deemed to be reasonable restrictions.

CHAPTER IV POWER TO MAKE ORDERS DETAINING CERTAIN PERSONS

8. Detention of certain persons.

(1) The Government may -

(a) if satisfied with respect to any person that with a view to preventing him from acting in any manner prejudicial to

(i) the security of the State or the maintenance of the public order, or

(ii) [deleted by J & K Act XII of 1988]

(b) if satisfied with respect to any person who is -

(i) a foreigner within the meaning of the Foreigners Act; 1946, or

(ii) a person residing in the area of the State under the occupation of Pakistan, that with a view to regulating his continued presence in the state or with a view to making arrangements for his expulsion from the State, it is necessary so to do, make an order directing that such person be detained.

(2) any of the following officers, namely -

(i)Divisional Commissioners,

(ii)District Magistrates, may, if satisfied as provided in sub-clauses (i) and (ii) of clause (a) of sub-section (1), exercise the powers conferred by the said sub-section.

(3)For the purposes of sub-section (1), -

(a)[deleted by J & K Act XII of 1988]

(b)"acting in any manner prejudicial to the maintenance of public order" means (i)promoting, propagating, or attempting to create, feelings of enmity or hatred or disharmony on grounds of religion, race, caste, community, or region;

(ii)making preparations for using, or attempting its use, or using, or instigating, inciting, or otherwise abetting the use of force where such preparation, using, attempting, instigating, inciting. provoking or abetting, disturbs or is likely to disturb public order;

(iii)attempting to commit, or committing. or instigating, inciting, provoking or otherwise abetting the commission of mischief within the meaning of section 425 of the Ranbir Penal Code where the commission of such mischief disturbs, or is likely to disturb public order;

(iv)attempting to commit, or committing, or instigating, inciting, provoking or otherwise abetting the commission of an offence punishable with death or imprisonment for life or imprisonment for a term extending to seven years or more, where the commission of such offence disturbs, or is likely to disturb public order.

(4)When any order is made under this section by an officer mentioned in sub-section (2), he shall forthwith report the fact to the Government together with the grounds on which the order has been made and such other particulars as in his opinion have a bearing on the matter, and no such order shall remain in force for more than twelve days after the making thereof unless in the meantime it has been approved by the Government.

9. Execution of detention orders.

A detention order may be executed at any place in the manner provided for the execution of warrants of arrest under the Code.

10. Power to regulate place and conditions of detention.

Any person in respect of whom a detention order has been made under section 8 shall be liable -

(a)to be detained in such place and under such conditions including conditions as to the maintenance of discipline and punishment for breaches of discipline as the Government may, by general or, special order, specify and,

(b)to be removed from one place of detention to another place of detention in the State by order of the Government.

11. Detention orders not to be invalid or inoperative, on certain grounds.

No detention order shall be invalid or inoperative merely on the ground -

(a)that the person to be detained thereunder is outside the limits of the territorial jurisdiction of the officer making the order; or

(b)that the place of detention of such person is outside the said limits.

12. Powers in relation to absconding persons.

If the Government, or an officer specified in sub-section (2) of section 8, as the case may be, has reason to believe that a person in respect of whom a detention order has been made has absconded or is concealing himself so that order cannot be executed, the Government or the officer may -

(a)make a report in writing of the fact to a Magistrate of the first class having jurisdiction in the place where the said person ordinarily resides, and thereupon the provisions of sections 87, 88 and 89 of the Code shall apply in respect of the said person and his property as if the order directing that be detained were a warrant issued by the Magistrate

(b)by notified order direct the said person to appear before such officer, at such place and within such period as may be specified in the order, and if the person fails to comply with such direction, he shall, unless he proves that it was not possible for him to comply therewith and that he had within the period specified in the order, informed the officer mentioned in the order of the reason which rendered compliance therewith impossible and of his whereabouts, be punishable with imprisonment for a term which may extend to one year or with fine or with both.

13. Grounds of order of detention to be disclosed to persons affected by the order.

(1) When a person is detained in pursuance of a detention order, the authority making the order shall, as soon as may be but ordinarily not later than five days and in exceptional circumstances and for reasons to be recorded in writing, not later than ten days from the date of detention communicate to him the grounds on which the order has been made, and shall afford him the earliest opportunity of making a representation, against the order to the Government.

(2) Nothing in sub-section (1) shall require the authority to disclose facts which it considers to be against the public interest to disclose.

14. Constitution of Advisory Board.

(1)The Government shall whenever necessary, constitute an Advisory Board for the purposes of this Act.

(2)Such Board shall consist of a Chairman, who is or has been a Judge of the High Court, and two other members who are, or have been, or are qualified to be appointed as Judges of the High Court.

(3)The Chairman and the other members of the Board shall be appointed by the Government in consultation with the Chief Justice of the High Court.

15. Reference to Advisory Board.

In every case where a detention order has been made under this Act, the Government shall, within four weeks from the date of detention order, place before the Advisory Board constituted by it under section 14, the grounds on which the order has been made, the representation, if any, made by the person affected by the order and in case where the order as been made by an officer, also the report by such officer under subsection (4) of section 8.

16. Procedure of Advisory Board.

(1)The Advisory Board shall, after considering the material placed before it and, after calling for such further information as it may deem necessary from the Government or from the person concerned and if in any particular case it considers it essential to do so or, if the person concerned desires to be heard, after hearing him in person, submit its report to the Government within eight weeks from the date of detention.

(2)Notwithstanding anything contained in sub-section (1), the Board may, if the person detained so demands, at any time before submitting its report, after affording an opportunity to the person detained and the Government or the officer, as the case may be, of being heard, determine whether the disclosure of facts, not disclosed under sub-section (2) of section 13 to the person detained, is or is not against public interest. Such finding of the Board shall be binding on the Government.

(3)The report of the Advisory Board shall specify in a separate part thereof the opinion of the Advisory Board as to whether or not there is sufficient cause for the detention of the person concerned.

(4)When there is a difference of opinion among the members forming the Advisory Board, the opinion of the majority of such members shall be deemed to be the opinion of the Board.

(5) Nothing in this section shall entitle any person against whom a detention order has been made to appear by any legal practitioner in any matter connected with the reference to the Advisory Board and the proceedings of the Advisory Board and its report, excepting that part of the report in which the opinion of the Advisory Board is specified, shall be confidential.

17. Action upon the report of Advisory Board.

(1)In any case where the Advisory Board has reported that there is in its opinion sufficient cause for the detention of a person, the

Government may confirm the detention order and continue the detention of the person concerned for such period as it thinks fit.

(2)In any case where the Advisory Board has reported that there is, in its opinion, no sufficient cause for the detention of the person concerned, the Government shall revoke the detention order and cause the person to be released forthwith.

18. Maximum period of detention.

(1)The maximum period for which any person may be detained in pursuance of any detention order which has been confirmed under section 17, shall be -

(a)twelve months from the date of detention in the case of persons acting in any manner prejudicial to the maintenance of public order or indulging in smuggling of timber; and

(b)two years from the date of detention in the case of persons acting in any manner prejudicial to the security of the State.

(2)Nothing contained in this section shall affect the powers of the Government to revoke or modify the detention order at any earlier time, or extend the period of detention of a foreigner in case his expulsion from the State has not been made possible.

19. Revocation of detention orders.

(1)Without prejudice to the provisions of section 21 of the General Clauses Act, Smvat 1977, a detention order may, at any time, be revoked or modified by the Government, notwithstanding that the order has been made by any officer mentioned in sub- section (2) of section 8.

(2)There shall he no bar to making of a fresh order of detention against a person on the same facts as an earlier order of detention made against such person in any case where -

(i)the earlier order of detention or its continuance is not legal on account of any technical defect or

(ii)the earlier order of detention has been revoked by reason of any apprehension, or for avoiding any challenge that such order or its continuance is not legal on account of any technical defect

Provided that in computing the maximum period for which a person against whom such fresh order of detention has been issued may be detained, the period during which such person was under detention under the earlier order of detention shall be excluded.

20. Temporary release of persons detained.

1. The Government may, at any time, order that a person detained in pursuance of detention order may be released for any specified period either without conditions or upon such conditions specified in the direction as that person accepts and may, at any time, cancel his release.
2. In directing the release of any person under sub-section (1), the Government may require him to enter into a bond with or without sureties for the due observance of the conditions specified in the direction.
3. Any person released under sub-section (1), shall surrender himself at the time and place and to the authority, specified in the order directing his release or cancelling his release as the case may be.
4. If any person fails without sufficient cause to surrender himself in the manner specified in sub-section (3) be shall be punishable with imprisonment for a term which may extend to two years or with fine, or with both.

5. If any person released under subsection (1) fails to fulfil any of the conditions imposed upon him under the said sub-section or in. the bond entered into by him, the bond shall be declared to be forfeited and any person bound thereby shall be liable to the penalty thereof.
6. The period of release shall not count towards the total period of detention undergone by the person released under this section.

CHAPTER V MISCELLANEOUS

21. Cognizance of offences under this Act.

1) No Court shall take cognizance of any offence under this Act except on a report in writing made by a public servant.
2) Notwithstanding anything contained in the Second Schedule to the Code, offences under this Act shall be cognizable and non-bailable.

22. Protection of action taken under this Act.

No suit, prosecution or any other legal proceeding shall lie against any person for anything done or intended to be done in good faith in pursuance of the provisions of this Act.

23. Power to make rules.

The Government may, by notification, make such rules consistent with the provisions of this Act, as may be necessary for carrying out the objects of this Act.

24. Repeal and Saving.

(1) The Jammu and Kashmir Public Safety Ordinance, 1977 is hereby repealed.

(2) Not withstanding such repeal, anything done or any action taken (including any rule or order made) under the said Ordinance shall, so far as consistent with the provisions of this Act, be deemed to have been done or taken under the corresponding provisions of this Act

.

References

Acharya, A. (2001), "Human Security: East versus West", *International Journal*, 56 (3): 442-460.

Acharya, A. and Acharya, A. (2001), Human Security in Asia: Conceptual Ambiguities and Common Understandings. *Paper presented in Centre for Peace and Development Studies*, Orissa: India, with York University.

Acharya, A. and Christie R. (2008), "Human Security Research: Progress, Limitations and New Directions", Working Paper No. 11-08, Report of a Conference Organised by the Centre for Governance and International Affairs, Department of Politics University of Bristol, 8-11 February.

Acharya, A. et al. (2011). *Human Security: From Concept to Practice Case studies from Northeast India and Orissa*. USA: World Scientific Publishing.

Adhikari, S. and Kamle, M. (2010), "The Kashmir: An Unresolved Dispute between India and Pakistan", *Geopolitics Quarterly*, 6 (4): 58-107.

Ahamad, F. and E. Yasmin (2012), "Impact of turmoil on the handicraft sector of Jammu and Kashmir: An economic analysis", *International NGO Journal*, 7 (5): 78-83.

Ahmad, S. (2013), *Impact of Militarization on Education in Kashmir*, Working Group on Peace, Conflict and Education, Teachers College: Columbia University.

Al Jazeera (2010), "Kashmir Killings Proke Protesters", [Online: Web] Accessed on 27 Sept. 2013

Alkire, S. (2003), "A Conceptual Framework for Human Security", *Centre for Research on Inequality, Human Security and Ethnicity*, Working Paper-2, London: University of Oxford.

Amin, S. A. (2003), "A Re-evaluation of the Kashmir Dispute", Pakistan Horizon, 56 (2): 37-51.

Amin, S. and A. W. Khan (2009), "Life in conflict: Characteristics of Depression in Kashmir", *International Journal of Health Sciences*, 3 (2): 213-223.

Amnesty International (2011), *India: A 'Lawless Law': Detentions under the Jammu and Kashmir Public Safety Act*, Amnesty International Ltd., London.

Anderson J. B (1991), "An International Criminal Court-An Emerging Idea", *Nova Law Review* 5 (3): 433-447.

Anderson, M. B (1996), *Humanitarian NGOs in Conflict Intervention in Managing Global Chaos*. in Crocker, Hampson, F & Hall, P. (eds.), Washington, DC: United States Institute of Peace Press.

Annan, K. (2000), "United Nations Millennium Report", 12 September, URL: http://www.un.org/ millenium/sg/report/full.htm.

Arie, S. et.al. (2000), "Human Security: A Framework for Assessment in Conflict and Transition", Project Report, United States Agency for International Development (USAID) Complex Emergency Response and Transition Initiative, New Orleans, LA: Tulane University.

Arora, V. and Dhingra, R. (2005), "At the Cross Roads: Families in Distress" *J. Hum. Ecol.,* 17(3): 217-222.

Arshad, H. (2008), "*Education in Kashmir: Historical Overview and Current Analysis*", Kashmir Corps, Germantown.

Asia Watch and Physicians for Human Rights (1993), *Pain in Kashmir: A Crime of War.*

Aurobinda, D. and S. Shekhawat (2008), "The Peace Process and Prospects for Economic Reconstruction in Kashmir", *Peace & Conflict Review,* 3 (1): 1-17.

Axworthy, L. (1997), "Canada and Human Security: The Need for Leadership", *International Journal,* 52 (2): 183-196.

Baba, N.A. (2012), *Kashmir Special Status: Myth and Reality.*in Raghavan, V.R (eds.) *Conflict in J and K: Impact on Polity, Economy and Society*, New Delhi: VIJ Books India Pvt. LTD.

Bajpai, K. (2000), "Human Security: Concept and Measurement," *Kroc Institute* ***for International Peace Studies*** Occasional Paper No. 19, USA.

Bakaya, P and Bhatti, S. (2005), "Kashmir Conflict: A Study of What Led to the Insurgency in Kashmir Valley and Proposed Future Solutions"
URL:http://www.stanford.edu/class/e297a/Kashmir%20Conflict%20%20A%20Study%20of%20What%20Led%20to%20the%20Insurgency%20in%20Kashmir%20Valley.pdf.

Bamzai, P.N.K. (1994), *Culture and Political History of Kashmir*, New Delhi: M .D. Publications Pvt Ltd.

Basrur, R. M. (2007), *Introduction: Bringing Human Security Back into South Asia.* In Basrur, R.M., and Joseph, M. (eds.) *Reintroducing the Human Security Debate in South Asia*, New Delhi: Institute of Peace and Conflict Studies.

BBC News (2008), "Police Fire on Kashmir Protesters", [Online: Web] Accessed on 19 Sept. 2013.

Behera, N. C (2000), *State, Identity, and Violence: Jammu, Kashmir, and Ladakh*, New Delhi: Manohar Publications.

Bose, S. M (2003), *Kashmir: Roots of Conflict, Paths to Peace,* New Delhi: Vistaar Publications.

Boutros, G. (1992), An Agenda for Peace, Preventive diplomacy, peace-making and peace-keeping accessed at http://www.cfr.org/peacekeeping/report-un-secretary-general-agenda-peace/p23439.

Bracken, Paul (1999) *Fire in the East: The Rise of Asian Military Power and the Second Nuclear Age,* New York: Harper Collins.

Breacher, M. (1953), *The Struggle for Kashmir*, New York: Oxford University Press.

Brown, W.N. (1996), *The United States, India and Pakistan*, Cambridge, MA: Harvard UP.

Burchill, S. (2001), "Liberalism", in S, Burchill et al (eds), *Theories of International Relations*, UK: Palgrave Macmillan.

Chennoy, A.M., and S. Tadjbakhsh (2007), *Human Security: Concepts and Implications*, New York: Routledge.

Cohen, S. P. (2002), "India, Pakistan and Kashmir", *Journal of Strategic Studies,* 25 (4): 32-60.

Cornelis, E.A. (2012), "Japan and the Human Security Debate: History, Norms and Pro-active Foreign Policy", *Graduate Journal of Asia-Pacific Studies,* 3(2): 58-74.

Das, T. (1950), "The Kashmir Issue and the United Nations", *Political Science Quarterly*, 65 (2): 264-282.

Datta, D. et al. (2008), "The Peace Process and Prospects for Economic Reconstruction in Kashmir", *Peace & Conflict Review*, 3(1): 1-17.

Donnelly, J. (1993), *International Human Rights*, Boulder Colo: Western Press.

Doxey, M. (1996), *International Sanctions in Contemporary perspective*, London: Macmillan.

Duschinski, H. (2009), "Destiny Effects: Militarization, State Power, and Punitive Containment in Kashmir Valley", *Anthropological Quarterly*, 82(3): 691.

Economic Times, (2013), "Four Killed in Kashmir Protests against Alleged Quran Desecration", [Online: Web] Accessed on 24 Sept. 2013.

Evans, A. (2000). "The Kashmir Insurgency: As Bad as it Gets", *Small Wars Insurgencies* 11(1): 69.

Fayyaz, A. A. (2012), "Top Army, police officials involved in human rights abuse in Kashmir". *The Hindu*, 7 December, 2012.

Floyd, R. (2007), "Human Security and the Copenhagen School's Securitization Approach: Conceptualizing Human Security as a Securitizing Move", *Human Security Journal* 5: 38-49.

Free Press Kashmir, (2008), "Unrest: Security Forces Violated SOP's during Protests", [Online: Web] Accessed on 13 Sept. 2013.

Ganguly, S. (1995), "Wars without End: The Indo-Pakistani Conflict", *Annals of the American Academy of Political and Social Science,* 541: 167-178.

Ganguly, S. & Bajpai, K. (1994), "India and the Crisis in Kashmir", *Asian Survey*, 34 (5): 401-416.

Ganguly, S. (1990), "Avoiding War in Kashmir", *Foreign Affairs*, 69 (5): 57-73.

Ganguly, S. (2003), "Explaining the Kashmir Insurgency, Political Mobilization and Institutional Decay", *International Security,* 21 (2): 76-107.

Ganguly, S. (ed) (2003), *The Kashmir Question: Retrospect and Prospect*, New York: Frank Class.

Gasper, D. (2008), "The idea of Human Security", *Institute for Social Studies*, Working Paper: No 28/08, The Hague.

Gayathri, A. (2012), "India Ignoring Rights Abuses In Kashmir", *International Business Times*, 07 December, 2012.

Ghosh, P. (2013), "Kunan-Poshpora: Can Kashmiri Rape Victims get Justice against the Indian Military after two Decades", International Business Times, Srinagar, 5 August 2013.

Gomez, Oscar A (2012) 'What is a human security project? The experience of the UN Trust Fund for Human Security', *Peace, Security & Global Change*, 24 (3): 385-403.

Government of Jammu and Kashmir (2012-13), *Economic Survey,* Directorate of Economics and Statistics, J and K.

Government of Sweden (1991), *The Stockholm Incentive on Global Security and Governance*, Prime Minister Office, Stockholm.

Gul, K (2013), "Shopian: Curfew on 13th day", *Greater Kashmir,* Srinagar, 20 September 2013.

Gupta, B. *S* (1970), *The Fulcrum of Asia*: Relations Among *China*, India, Pakistan, and the U.S.S.R., New York: Pegasus.

Habibullah, W. (2008), *My Kashmir: Conflict and the Prospects of Enduring Peace*, Washington DC: United States Institute of peace.

Hajni, M. (2007-08), "The Kashmir Conflict: A Kashmiri Perspective", *Swords and Ploughshares,* XVI (1): 12-14.

Hampson, F. O., et al. (2002), *Madness in the Multitude: Human Security and World Disorder,* Toronto: Oxford University Press.

Hassan, A and Shafi A. (2013), "Impact of Conflict Situation on Mental Health in Srinagar, Kashmir", *Bangladesh e-Journal of Sociology*, 10 (1): 101-126.

Hassan, F. (2016) "World Bank ranks JK among 'poorest' performers in India", *The Kashmir Monitor*, Srinagar, 8 November 2016.

Henk, D. (2005), "Human Security: Relevance and Implications", *Parameters: US Army War College Quarterly*, 35 (2): 91-106.

Heymann, K. (2003), "Earned Sovereignty for Kashmir: The Legal Methodology to Avoiding a Nuclear Holocaust," *American University International Law Review,* 19 (1): 152-200.

Hilali, A.Z. (2007), "Kashmir dispute and UN mediation efforts: An historical perspective", *Small Wars and Insurgencies*, 8 (2): 61-86.

Hoffmana, B. and Duschinski, H. (2013), *Contestations over Law, Power and Representation in Kashmir Valley*, USA: Routledge.

Huliaras, A. and Tzifakis, N. (2007), "Contextual Approaches to Human Security Canada and Japan in the Balkans," *International Journal,* 62 (3): 559-575.

Human Rights Watch and Physicians for Human Rights (1993), *The Human Rights Crisis in Kashmir: A Pattern of Impunity.*

Hussain, S.R. (2009), "Resolving the Kashmir Dispute: Blending Realism with Justice," *Pakistan Development Review,* 48 (4): 1007–1035.

Hussein, Karim. Gnisci, Donata and Wanjiru, Julia (2004) *Security and Human Security: An Overview of Concepts and Initiatives; what Implications for West Africa?*, OECD.

ICG Asia Report (2003), *Kashmir: The View from New Delhi,* N°69 New Delhi/Brussels.

Independent Commission on International Development Issues (1980), *North-South: A programme for survival*, Cambridge, M.A: MIT Press.

Independent Peoples Tribunal (2010), *Report of Independent People's Tribunal on Human Rights Violations in Kashmir*, Human Rights Law Network, New Delhi.

Islam, Ajaz Ul (2014), "Impact of Armed Conflict on Economy and Tourism: A Study of State of Jammu and Kashmir," *Journal of Economics and Finance,* 4 (6): 55-60.

James, C.C. and Zdamar, O. (2005), "Religion as a Factor in Ethnic Conflict: Kashmir and Indian Foreign Policy", *Taylor & Francis,* 17: 447–467.

Jolly, R. and Ray, D.B. (2006), The Human Security Framework and National Human Development Reports: A Review of Experiences and Current Debates. *National Human Development Report Unit.* Institute of Development Studies, Sussex.

Jones, S. (2008), "India, Pakistan, and counterinsurgency operations in Jammu and Kashmir," *Small Wars and Insurgencies*, 19 (1): 1-22.

Jones, Z., Singletary, C. et.al (2010), *Economic Development as a Tool to Reduce Secessionism in Jammu and Kashmir.* Spring: University of Wisconsin Madison.

Jong, K., Saskia, V.K. (et al) (2006), "Kashmir: Violence and Health", *Medecins Sans Frontiers*, Amsterdam.

Kathju, J. (2013), "Over 2 Lakh Kashmiri Youth Registered in 3 Months," *Rising Kashmir,* Srinagar, 24 November 2013.

Kazi. S (2007), *Between Democracy and Nation: Gender and Militarisation in Kashmir*, Ph. D. Thesis, UK: London School of Economics and Political Science- The Gender Institute.

Khan, B. A. (2013), "Voilences and Turmoils Adversely Effected Jammu and Kashmir Economy", *International Journal of Educational Research and Technology,* 4 (1): 18-21.

Khan, B. A. and Kiyawat, B. (2013), "Inclusive Growth in Jammu and Kashmir: Problems and Prospects", *International Journal of Science and Research (IJSR)*, 2 (5): 238-241.

Kis, A. (2010), "The Operationalization of the Human Security Concept," *Revista Română de Geografie Politică*, 12 (2): 254-266.

Krause, K. (2007), "Towards a Practical Human Security Agenda," Policy Paper, No. 26, Geneva Centre for the Democratic Control of Armed Forces (DCAF).

Kymlicka, W. (1996), *Multiculturalism Citizenship: A Liberal Theory of Minority Rights*, UK: Oxford University Press.

Lamb, A. (1996), *The Kashmir Problem*, New York: Praeger

Lamb, A. (1997), Incomplete Partition: The Genesis of the Kashmir Dispute, 1947-1948, London: Oxford University Press.

Lauren, P.G. (1998), *The Evolution of Human International Rights: Visions Seen.* University Park: Pennsylvania State University Press.

Mahapatra, D. and Shekhawat, (2008), *Kashmir across LOC*, New Delhi: Gyan Publishing House.

Malhotra, C. (2007), "Internally Displaced People from Kashmir," *Indian Anthropologist*, 37 (2): 71-80.

Malik, I. (2002), *Kashmir: Ethnic Conflict and International Dispute.* London: Oxford University.

Marwah, V. (2007), "Jammu and Kashmir / India", Kreddha Autonomy Mapping Project, a project of the Kreddha research and analysis, *International Peace Council for States and Minorities*, 1 – 21 October.

McGivering, Jill (2000), "Kashmir's Orphaned Thousands," BBC News, Srinagar, 7 August 2000.

Minami, R. (2006), "Human Security and Japan's Foreign Policy,". *Gaiko Forum, Japanese Perspectives on Foreign Affairs*, 5 (4): 43-51.

Mohiuddin, L. (1997), "Human Rights Violations: A Case Study of Kashmir," *Pakistan Horizon.*

Moore, J. (1996), *The UN and Complex Emergency: Rehabilitation in the Third World Geneva*, UNRISD (United Nations Research Institute for Social Development).

Moravcik, A. (1995), "Explaining International Human Rights Regimes: Liberal Theory," *European Journal of International Relations* 1 (2): 157-189.

Navlakha, B. (1991), "Bharat's Kashmir War", *Economic and Political Weekly,* 26 (51): 2951-2962.

Navlakha, G. (2008), "Resentment Persists in Kashmir", *Economic and Political Weekly*, 43 (6): 13-15.

Newman, E. (2001), "Human Security and Constructivism", *International Studies Perspectives,* 2: 239-251.

Noorani, A.G. (1994), "Kashmir and a Farcical Human Rights Commission", *Economic and Political Weekly*, 29 (32): 2053.

Noorani, A.G. (2000), "Kashmir and National Human Rights Commission", *Economic and Political Weekly*, 35 (21/22): 1785-1787.

Paris, R. (2001), "Human Security: Paradigm Shift or Hot Air?," *International Security*, 26 (2): 87-102.

Parvaiz, A. (2007), "School enrolment in J&K", *Kashmir Times*, Srinagar, 7 July 2007.

Prezelj, I. (2008), "Challenges in Conceptualizing and Providing Human Security", *HUMSEC Journal*, 1 (2): 1-22.

Puri, B. (1999), "Kashmir Problem Thrives on Denial of Human and Democratic Rights", *Economic and Political Weekly*, 34 (14): 794-795.

Qadir, S. (2009), "An Analysis of the Kargil Conflict 1999", *The Pakistan Development,* 48 (4): 1007–1035.

Qutab, S. (2012), "Women Victims of Armed Conflict: Half-widows in Jammu and Kashmir," *Sociological Bulletin*, 61 (2): 255-278.

Raj, M. (2010), "Kashmir Crisis due to Unemployment," *The Kashmir Times,* Srinagar, 22 July 2013.

Rajagopalan, R. (2012), "*Security*", in Chimni, B.S. and Mallavarapa, S. *International Relations: Perspectives of the Global South.* USA: Pearson.

Rather, F.A. (2013), "Armed Conflicts in J&K and Its Impact on Society: A Case Study of Kashmir Valley," *International Journal of Scientific and Research Publications*, 3 (2): 1-3.

Rather, M. A. 2017. "Kashmir Conflict: From Arms Struggle to Peaceful Protests," *Rising Kashmir*, 25 May, 2017

Rehman, Ejaz U (2009), *Impact of Armed Conflict on Environment in the State of Jammu and Kashmir*, URL: http://www.scribd.com/doc/14892353/ Impact-of-Armed-Conflict-on-Environment-in-the-State-of-Jammu-and-Kashmir#scribd.

Remacle, E. (2008), "Approaches to Human Security: Canada, Japan and Europe in Comparative Perspective", *The Journal of Social Science,* 66: 5-34.

Schofield, V. (2000), *Kashmir in Conflict: India, Pakistan and the Unending War.* London: I.B.Tauris & Co Ltd.

Shad, D. (2013), "Impact of Armed Conflict on Kashmir Tourism and Socio-Economic Profile of the People of J&K", *EXCEL International Journal of Multidisciplinary Management Studies*, 3 (7): 35-54.

Shah, Farooq A. Wani, Ambreen and Mustafa, Zuhaib (2013) "Impact of Armed Conflict on Kashmir Tourism and Socio-Economic Profile of the People of J and K", *International*

Shekhawat, S. (2006), *Conflict Induced Displacement: The Pandits of Kashmir,* New Delhi: Saksham Books International.

Shekhawat, S. (2009), "Fragile Kashmir, Costs and Hopes for Peace." *Journal of Alternative Perspectives in the Social Sciences* 1 (3): 976-981.

Shinoda, H. (2004), "The Concept of Human Security: Historical and Theoretical Implications." *In*: Shinoda, H and Jeong, H.W. (eds.), *Conflict and Human Security: A Search for New Approaches of Peace-building.* Institute for Peace Science: Hiroshima University. IPSHU English Research Report Series, No.19.

Siddhartha, P. (2000), "Political Economy of Kashmir since 1947", *Economic and Political Weekly*, 35 (24): 2051-2060.

Singh, A. (2011), *Conflict in Jammu and Kashmir*. Bangalore: National Institute of Advance Studies.

Singh, P. (1995), An Indian Perspective II, *Asian Affairs* 22 (1): 40-53.

Smita, J. (2003), *Kashmir History, Current Issues, and a Proposal for a Brighter Future* accessed at http://www.stanford.edu/class/e297a/india%20% 20kashmir%20%20 pakistan. htm.

Suzuki, T. (2012), "Japan As a Champion of Human Security." *Global Health Governance*, 6(1): 1-3.

Tabassum, M. T. (2012), "Political Situation in Kashmir and Role of United Nations." SCS *Journal*, 1 (2): 4-28.

Tajbakhsh, S. (2005), *Human Security: Concepts and Implications with an Application to Post-Intervention Challenges in Afghanistan*: New Delhi: Centre for Peace and Conflict Resolution.

Takasu, Y. Statement by Mr. Yukio Takasu, Director-General of multilateral co-operation department at the Third Intellectual Dialogue on Building Asia's tomorrow, Bangkok, 19 June 2000, accessed at http://www.mofa.go.jp/policy/ human_secu/speech0006.html on 18 September 2013).

Teng, M.K. And Gadoo, C.L. (1991), *White Papers on Kashmir*. New Delhi: Geoffrey Bell and Publisher.

The Hindu, (2011), Turmoil has taken a toll on psychological health of Kashmiris, says NGO report, [Online: Web] Accessed 20 Nov. 2013.

The Tribune (2008), "Dropout rate down in schools - Jammu & Kashmir," Chandigarh, 26 Feb 2008.

Thorner, A. (1949), "The Kashmir Conflict," *Middle East Journal*, 3 (1): 17-30.

***Times of India*, (2013), J and K: C Strike Disrupts Normal Life in Kashmir, Curfew on in Shopian, [Online: Web] Accessed 25 Sept. 2013.**

Tsai, Y. (2009), "The Emergence of Human Security: A Constructivist View", *International Journal of Peace Studies*, 14 (2): 19-33.

United Nations Development Programme (1994), *Human Development Report 1994,* New York: Oxford University Press.

United Nations Development Programme (1997), *Human Development Report 1997,* New York: Oxford University Press.

United Nations Document Official Report (1948), New York: Oxford University Press.

United Nations Report of the World Commission on Environment and Development (1987), *Our Common Future*, New York: Oxford University Press.

Varshney, A. (1991), "India, Pakistan, and Kashmir: Antinomies of Nationalism" *Asian Survey*, 31 (11): 997-1019.

Wani, H. A. and Suwirta, A. (2013), "Understanding Kashmir Conflict: Looking for its Resolution," *Susurgalur*, 1 (2): 179-192.

Wani, M. Z. and Shah, F.A. (2013), "Impact of Armed Conflict on Kashmir Tourism and Socio-Economic Profile of the People of J and K", *International Journal of Multidisciplinary Management Studies (EXCEL),* 3 (7): 35-44.

www.ingramcontent.com/pod-product-compliance
Lightning Source LLC
LaVergne TN
LVHW051050220926
848485LV00001BA/60
* 9 7 8 1 6 8 0 5 3 4 7 2 6 *